To Dad

love

Sheena + John

XX

KU-152-050

ANIMALWATCHING

DESMOND MORRIS

by the same author

The Biology of Art
Men and Snakes (*co-author*)
Men and Apes (*co-author*)
Men and Pandas (*co-author*)
The Mammals: A Guide to the Living Species
Primate Ethology (*editor*)
The Naked Ape
The Human Zoo
Patterns of Reproductive Behaviour
Intimate Behaviour
Manwatching
Gestures (*co-author*)
Animal Days
The Soccer Tribe
Inrock (*fiction*)
The Book of Ages
The Art of Ancient Cyprus
Bodywatching
The Illustrated Naked Ape
Dogwatching
Catwatching
The Secret Surrealist
Catlore
The Animals Roadshow
The Human Nestbuilders
Horsewatching

JONATHAN CAPE · LONDON

ANIMALWATCHING

A Field Guide to Animal Behaviour

DESMOND MORRIS

Editor: Liz Smith
Designer: Ian Craig
Picture Researcher: Dee Robinson
First published 1990
© Desmond Morris 1990
Jonathan Cape Ltd,
20 Vauxhall Bridge Road, London SW1V 2SA
A CIP catalogue record for this book is available from the British Library
ISBN 0-224-02794-8
Phototypeset by Keyspools Ltd, Golborne, Lancashire
Printed in Germany by
Mohndruck GmbH, Gütersloh

Picture Credits

Aquila: pp 50 centre; 132 upper; 134 upper, centre and lower left; 221 upper; 227 upper; 231; 236 upper; 238 upper and centre.
Ardea: pp 5 lower; 6; 7 upper; 18 upper right, lower left and right; 22 lower; 25 lower; 26; 28 lower left; 35 upper; 36 lower right; 37; 41; 46 right; 49 lower; 56 right; 66 lower; 68; 70 lower; 73; 77 upper; 88 left; 90 upper; 100; 107 lower; 120 upper and lower; 123; 125 upper; 130 upper; 146 upper; 150 lower left; 153 lower; 161 left and right; 162 upper and lower right; 169 upper; 172 lower; 184 right; 186 lower; 188; 199 lower right; 208; 212 left; 218 upper right and lower left; 223 top right; 229; 234 upper; 238 lower; 240 upper, centre and lower; 241 upper; 243 upper; 250 right.
Biofotos: pp 23 upper; 52 lower; 71; 85 lower; 179 upper; 192 upper; 203; 206 upper; 212 right; 219; 223 upper right; 224 lower left; 229.
Chris Bowden: p 4.
The British Library: pp 82; 93.
Bruce Coleman: pp 13 upper and centre right, lower left; 15; 18 upper left; 22 upper; 32 left; 34; 36 lower left; 43 centre; 44; 45 upper; 52 centre; 55; 57 lower right; 72 lower; 77 lower; 80 right; 83 upper and lower; 84 left; 85 upper; 87 lower; 90 lower; 103 upper left; 109 upper and lower; 110; 111 upper and lower; 114 right; 124 upper; 127 upper and lower; 129 upper and lower; 135 upper; 138; 140 left and right; 142; 143 lower; 160 lower; 168 upper; 170 upper left and right; 184 left; 186 upper; 187 lower; 190 upper and lower left; 194 left and right; 195 upper left; 199 upper and lower left; 204 right; 205 left; 213 upper; 214 upper and lower left and right; 216 lower; 221 lower; 222 lower; 227 lower; 233 left; 237 upper left; 239; 241 lower left; 242 upper; 245 upper; 250 left.
Food and Agricultural Organisation of the UN: p 19 upper.
Eric and David Hosking: pp 84 right; 86 upper and lower; 92 upper, centre and lower left; 97 lower; 131 upper; 168 upper; 241 lower right.
Image Bank/Ocean Images: pp 48 upper and lower right; 119 lower; 125 lower.
Jacana: pp 25 upper; 47; 48 upper left; 50 lower right; 51 upper, centre and lower; 101 left; 103 lower left and right; 124 lower; 126 centre; 137 right; 144 upper and lower; 164 left and right; 173; 209 lower; 215 lower; 220 left; 223 left; 234 lower; 245 centre and lower.
Frank Lane Picture Agency: pp 23 lower; 81; 98 upper left; 147 lower; 150 upper left; 176 lower; 180 upper and lower; 193 lower right; 200–201; 209 upper; 226 upper left; 248 upper.
Mantis Wildlife: pp 5 upper; 7 lower; 32 right; 40 upper right; 49 upper; 50 lower left; 52 upper; 64 upper; 76 upper; 97 upper right; 146 lower; 190 right; 191; 199 upper right; 237 lower.
The Museum of Modern Art: p 12.
The National Audubon Society/Photo Researchers Inc: p 91 lower.
Trustees of the Natural History Museum: p 35 lower.
The National Photographic Index of Australian Wildlife: pp 128; 197 upper; 200 lower.
Natural History Photographic Agency: pp 5 centre; 8; 10 right; 28 upper; 28 lower right; 31; 36 upper; 40 upper left; 50 upper; 60 lower; 62; 69 upper; 70 upper; 72 upper right; 74 lower; 76 lower; 78 lower; 87 upper; 96 lower; 99 upper; 102 upper; 103 upper right; 104 right; 108 upper; 126 upper and lower; 145; 150 right; 152 upper and lower; 156; 163 lower; 168 lower; 172 upper; 174 right; 175 upper; 176 upper; 177; 197 lower; 198; 201 lower; 207 lower; 211; 226 upper right; 228 upper and lower; 232 upper right; 236 lower; 242 lower; 244 left and right.
Natural Science Photos: pp 13 upper left (P & S Ward) and centre left (K Cole); 16 left (A Watts); 21 upper; 24 lower; 40 lower (C Banks); 43 lower (I Bennett); 45 lower (C Mattison); 46 left (J Hobday); 57 upper and lower left (P & S Ward); 65 upper (D Hill); 66 upper left (D Hill); 75 upper and lower (P & S Ward); 78 upper and centre (P & S Ward); 95 (C Jones); 96 upper (K Cole); 97 upper left; 98 lower (C Jones); 105 left (P Kay); 112 (K Cole); 117 left (D Scott) and right (P & S Ward); 133 (L Hes); 137 left (P & S Ward); 139; 147 upper (O C Roura); 148 (A P Barnes); 153 upper (J Grant); 157 upper left (J Warden); 163 upper (A Smith); 167 (R Watson); 169 lower (B Gibbs); 178 upper left (B Gibbs); 181 upper (C Blancy) and lower (U Glimmerveen); 185 (D Yendall); 210 upper left (M Chinery); 224 upper left (M Chinery); 224 right (G Newlands); 246 upper right and lower (Lex Hes); 248 lower (C Jones).
Oxford Scientific Films: pp 10 left; 11; 20; 21 lower; 27 left and right; 29; 30 left; 33 lower; 42 upper and lower; 43 upper; 53 upper and lower; 54; 56 left; 57 upper right; 58 lower left; 59; 60 upper; 61; 74 upper; 79; 80 left; 88 right; 89 upper; 92 right; 94 upper; 98 upper right; 99 lower; 102 lower; 104 upper and lower left; 105 right; 106 lower; 108 lower; 115; 116; 118 left; 122 right; 132 lower; 149; 160 upper; 165 left and right; 175 lower; 178 upper right; 179 lower; 182 left and right; 192 lower; 193 upper right; 195 lower left and top right; 202; 204 upper and lower left; 205 lower right; 206 lower; 207 upper; 210 right; 216 upper; 217 upper and lower; 222 upper; 232 lower; 233 right; 243 lower; 246 upper left.
Planet Earth Pictures: pp 2–3; 16 right; 33 upper left; 38 lower left; 39; 58 upper and lower right; 63; 65 lower; 66 upper right; 67; 69 lower; 72 upper left; 89 lower; 101 centre; 106 upper; 107 upper; 114 left; 119 upper; 121; 122 upper left; 136; 141 left and right; 151; 157 lower left and right; 189 upper and lower; 195 lower right; 196; 205 upper right; 210 lower left; 220 right; 222 centre; 225 lower; 230; 232 upper left; 235 upper and lower; 248 centre; 251.
Scala: p 91 upper right.
Sothebys: p 91 upper left.
Spectrum Colour Library: p 1.
Survival Anglia: pp 17; 19 lower; 33 upper right; 38 upper; 64 lower; 94 lower; 101 right; 113; 118 right; 122 lower left; 130 lower; 131 lower; 134 right; 135 lower; 143 upper; 155 left; 162 left; 174 left; 178 lower; 183 left and right; 193 left; 213 lower; 215 upper left and right; 218 upper left; 223 lower right; 225 upper; 226 lower; 237 upper right; 247.
Tony Stone: pp 9; 14.
Zefa Picture Library: pp 30 upper and lower right; 154; 158; 187 upper.
Zoo Operations Ltd: pp 13 lower right; 38 lower right; 155 right; 166; 170 lower left; 171.

Contents

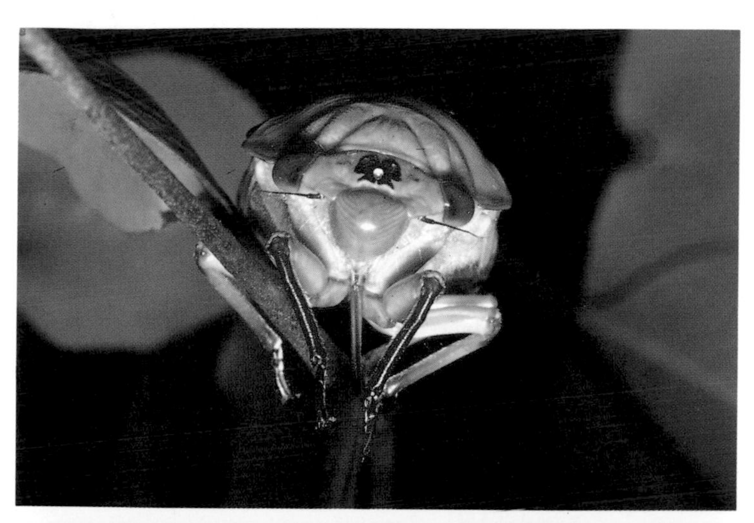

Introduction: The Animal-watcher

As a boy one of my greatest pleasures was watching animals. I divided my time between my hundreds of pet animals at home and the wild animals in the Wiltshire countryside. I spent so much time with them that I began to think like them, to see the world from their point of view. I identified with them so strongly that I began to see humans who hunted animals as the enemy. At the time this was an unfashionable attitude. Hunting-shooting-and-fishing was the norm. In the British countryside where I wandered, everyone did it. It was a way of life. But from an early age, for some reason, I rebelled. I preferred foxes to foxhounds. I found wildfowl more fascinating than wildfowlers. And I hated the anglers with their sharp hooks and their lack of understanding of what they were doing to the fish they plucked so gleefully from their wet world into our dry one.

Above all, I wanted to understand the world of animals. There were so many mysteries and it was hard to know how or where to begin. I tried to get closer to them. I discovered that sitting very still for as long as possible was the first great secret. Most people, I noticed, even experienced naturalists, were forever striding through the undergrowth, striking out across the fields, and searching, searching for something new. Far better, it seemed to me, was the simple strategy of waiting for nature to come to you, rather than going clumsily to look for it.

When Picasso was asked how he painted, he replied, 'I do not seek, I find.' Already, in my childhood, this had become my way of studying animals. When you walk into a wood or a field, you alarm everything that lives there. The moving human body is large, obtrusive and highly visible. But sit down quietly and, after a while, you become invisible. Nature resumes its activities, the patterns of behaviour you disrupted by your arrival. This is true whether you are in a desert or a forest or swimming on a coral reef.

When I was still very young I built myself a raft from old planks and oil-drums. I launched it on to a small lake and, lying flat on its wooden platform, pressed my face close to the water. The raft drifted very slowly, making no disturbance, and there through the mirror-smooth surface I saw a giant pike, lying in wait for its prey like a lurking U-boat. A shoal of young roach approached it unawares, sensed its presence and immediately closed ranks – safety in numbers – before darting off. I was so close to the water that I was already beginning mentally to enter their world and to feel their dramas as my dramas.

This was all happening half a century ago, before the invention of the aqualung. But I was already close enough to the aquatic world for it to become a lifelong obsession. It did not replace my fascination for mammals, birds and reptiles. It simply gave my study another dimension. My appetite for learning about all animals, simple or complex, was insatiable. Inevitably I was destined to become a zoologist in later life.

The great problem I faced, when I eventually obtained my degree in zoology, was that to convert my childhood fascination into an adult career I would have to carry out experiments on animals. Zoology was in an intensely experimental, laboratory-oriented phase, and this did not appeal to me. I was simply not prepared to treat animals in that way. In my mind I was one of them and there was no way that I was going to make a living from carrying out painful experiments on friends. It looked very much as though I would have to find some other career.

Just as I was about to give up zoology, I was lucky enough to attend a lecture by the great ethologist Niko Tinbergen. I had no idea what ethology was, but I soon found out. It was the naturalistic study of animal behaviour. Tinbergen demonstrated that it was possible, simply by watching animals, to make a scientific study of them. By making the watching systematic and analytical, it was possible to carry out field

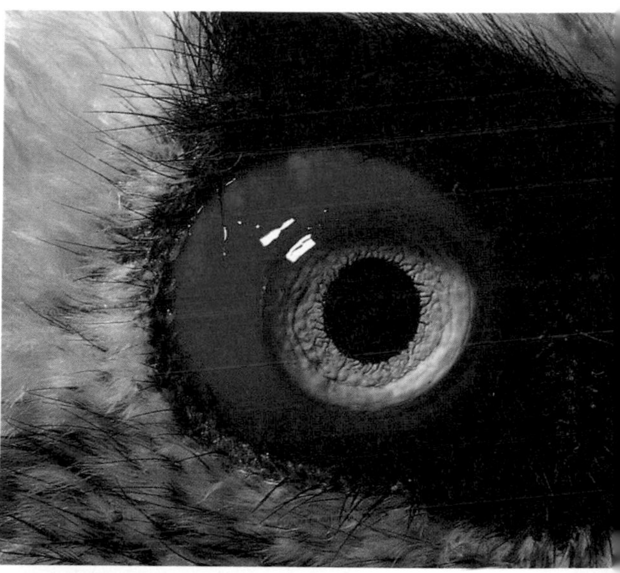

experiments that reduced interference with the animals to a minimum. He demonstrated that it was possible to convert amateur natural history into professional zoology by the straightforward device of quantified observation.

For me this was a revelation. It meant that, by carefully counting and scoring different animal actions in given time periods, I could make complicated analyses that helped to unravel the intricate behaviour patterns that existed in so many species. It meant that instead of guessing that the presence of a red spot or the flicking of a tail was acting as a threat signal or a courtship display, I could set about proving it. There was no looking back. A whole new world of animal study lay open to me and has continued to excite me ever since.

After writing about fifty scientific papers on the subject I decided to reach a wider audience and moved into television, where week after week I tried to get across my fascination for the animal world. My earlier studies of fish and birds were now extended to mammals and I ended up as the curator of the largest collection of mammalian species in the world – at the London Zoo. There, with a team of young research workers, I was able to extend my animal behaviour studies into areas not normally available to zoologists. I could work with everything from marmosets to elephants. I became especially interested in apes and made a long study of chimpanzees. It was at this point that I came face to face with the realisation that human beings could be studied like other animals, and I went on, past chimpanzees, to investigate the behaviour of this strange creature that I christened *The Naked Ape*.

I applied the same methods here, watching people rather than talking to

them, and I learnt many new things about the way humans interact. I published the results in a book called *Manwatching* and for twenty years continued my investigations into that bizarre species *Homo sapiens*.

Although my *Manwatching* study was based on my lifetime of animal-watching, I had never written the book of which it was, in effect, the sequel. At first I was too busy writing research reports and later on I had become completely engrossed in purely human studies. I eventually assumed that, by now, someone else must have written the animal version of *Manwatching* and recently I went to look for it. To my surprise it was not there. There were a number of excellent textbooks on the subject but nobody had taken the trouble to present scientific animal-watching to a general audience. And that is how the present volume came to be written.

Instead of providing discussions of theories and academic abstractions I have concentrated on animal behaviour patterns that can be seen and studied by any interested person. It helps a great deal to have a large number of clear, separate concepts, ideas that can quickly and easily be applied to the sometimes confusing variety of animal actions we observe when we sit down to watch other species. I have therefore made each chapter short, no more than a few pages, and have picked the most vivid examples in each case.

The serious student of animal behaviour starts out with a basic premise, namely that every spot of colour, every strange posture, every tiny movement that an animal makes, has some special meaning. Furthermore, all these colours and actions can be understood if they are studied closely enough. Nothing an animal ever does cannot be explained, given sufficient patience and ingenuity on the part of the animal-watcher. Everything has a reason, every piece of behaviour functions in some way to improve the chances of survival of the animal concerned. In the end, all animal mysteries can be unravelled.

To some romantics this may seem a pity. They would prefer to keep the mysteries of nature as mysterious as possible. They feel that to explain everything will be to destroy the beauty of nature, but they are wrong. To know that a particular animal dance, or brightly coloured display, operates as a territorial device or as a sexual arousal mechanism does not make it any less beautiful. To understand the function of bird-song does not make it any less enchanting. The romantic refusal to analyse is based on a fallacy. It also introduces a dangerous bias. For, to the romantic, the bird of paradise is much more exciting than the humble house sparrow. Any animal species that happens to be superficially dull will be ignored, perhaps even maltreated. But to the ethologist, every species is fascinating – the end-point of millions of years of complex evolutionary pressures. Every species has its own intricate behaviour repertoire of survival mechanisms and even the seemingly boring species soon become exciting when one starts to probe into their particular way of life. The social behaviour of the despised house sparrow is every bit as intriguing as that of the bird of paradise to the serious animal-watcher.

A final comment on my use of the word 'animal'. Many people confuse the terms 'animal' and 'mammal'. They speak of fish and insects, for example, as though they are not animals. What they mean, of course, is that they are not mammals. All living things are either plants or animals. The amoeba is just as much an animal as the elephant. So animal-watching covers everything from microscopic creatures to mammalian giants. But, having said this, I must admit that I have favoured familiar species or, at least, species that are relatives of familiar species. There is no point in explaining behaviour patterns in animals that are extremely difficult to observe. I have largely omitted the microscopic world, for example, as few people own microscopes. I have concentrated on the higher forms of life, not only because they are easy to observe but also to some extent because they have greater meaning for our own species. We can learn more about ourselves from monkeys than from microbes, and looking around our modern human world it is clear that we need to discover as much as we can about ourselves.

As I have said before, when writing about human behaviour, we are no more than swollen-headed animals and the sooner we accept this fact the safer we will be. As risen apes we have come to dominate this small planet so effectively that we are in danger of smothering it. There are many lessons we can learn from other animals, to our great and continuing advantage, and it is high time we took a little while to sit and stare at the other creatures with which we share the earth. They have much to teach us, as I hope the following pages will demonstrate . . .

Why does the Zebra have Stripes?

It comes as a shock to some people to learn that we still do not have definite answers to many of the most obvious questions about animals. Despite the enormous amount of information we have already accumulated there remains a great deal to be learned and this is one of the special pleasures of animal-watching. There is always the feeling that, at any moment, something may happen that will lead to a new discovery. And no matter how minor that discovery is, there is a peculiar excitement about understanding some natural phenomenon for the first time.

Frequently it is not the case that we have no answer to a question – rather, we have too many. Each answer seems to have some merit, but we cannot be certain which is the definitive one. What is needed is some new observation or insight to clarify the problem. An example will illustrate this.

Everyone knows that the zebra has bold black and white stripes, but why does it have them? What function do they serve and how do they help the zebra to survive? And what can we learn by a little careful zebra-watching?

The traditional explanation is that the stripes act as camouflage and help to break up the shape of the animal, concealing its body from the eyes of hungry predators. Despite their vivid appearance when seen in bright sunlight on open grassland, the zebras are said to be well camouflaged in broken cover at dawn and dusk, when predators are active and the light is dim. One experienced game tracker claimed that it was possible to get to within 40 or 50 yards before spotting a zebra under these conditions, and even then it only gave itself away by a small movement such as the swish of its tail or the sudden turn of its head. Under identical circumstances antelopes were visible up to 200 yards away.

This interpretation was accepted for many years but has recently been challenged. It has been pointed out that the behaviour of zebras simply does not match with this picture of them standing very still in broken cover. Compared with many hoofed animals on the plains of Africa, they are remarkably mobile and noisy and never attempt to hide in cover. On the contrary, they prefer to rest in groups out in the open, where they can use their acute senses to scan the landscape for tell-tale sounds, scents and movements. Their response to danger is to flee as fast as possible, rather than freeze or hide. In other words, their natural history does not fit with the explanation that has been repeated in textbooks for decade after decade. So some other answer must be sought.

The traditional view of zebra stripes is that they act as camouflage, but even in woodland cover this is not convincing and, in any case, zebra herds spend most of their time on open grassland, where the stripes cannot hide them. Some other function must be sought.

Certainly a visit to the plains of Africa gives one an immediate impression of how conspicuous the stripes are and how easy it is to spot a herd of zebra in almost any terrain. Perhaps, after all, they are meant to be seen. If so, why? One explanation is that they are meant to confuse a predator such as a lion – the zebra's chief enemy – by creating an optical illusion. If you compare a striped object with one that has a different kind of black and white pattern, they do not appear to be the same size even if, in reality, they are identical. If the striped pattern breaks against the outline of the object so that its contours are discontinuous, its shape is disrupted. For some strange optical reason this makes it look bigger than it really is and it has been argued that this illusion would have the effect of making an attacking lion misjudge its leap, striking short of the true position of the prey.

A close scrutiny of films showing lions killing zebras does not bear this out. When a lion gets close enough to strike, it usually does so with great skill and confidence and does not appear to be miscalculating its position to the slightest degree. If zebras escape, it is because of their initial alertness, then their speed and finally their stamina. It is the slow ones – the young, the old and the sick – that are caught, simply because they are not quick enough over a long enough distance.

A third suggestion is that the black and white stripes dazzle the lion when it gets very close to its prey and make it difficult for the killer to concentrate on its fast-moving victim. This is the Op Art Principle and anyone who has stood close to a painting by Bridget Riley will understand why dazzle can be a problem to sensitive eyes. The lines seem to shift and shimmer as you look at them and before long you have a powerful desire to look away, anywhere but at the painfully dazzling pattern in front of you.

In theory the dazzle explanation makes sense, because the zebra patterning is just about as vivid as it can be, when seen at close quarters, but once again the facts do not support this idea in practice. As with size-confusion, there is no evidence that dazzle-confusion puts off a hungry lion which is close enough to experience it.

It has been claimed that the vivid pattern of black and white stripes creates a dazzle effect, rather like the Op Art paintings of Bridget Riley, and that this disconcerts the attacking predator at close quarters. This is a possibility, but in practice there is no evidence for it.

Another suggestion is that the stripes confuse the lion by making it difficult to tell where one zebra ends and another begins. The herd tends to flee together and the stripes are thought to jumble up the individual shapes and make the fleeing herd look like one great mass of black and white patterning. Since lions always like to isolate one individual from the herd before attacking, this theory also makes sense on paper, but reality does not bear it out because lions can all too easily single out animals for the kill. This is what happens during a successful chase, as one individual falls behind or gets separated in some way from the rest.

Perhaps, then, the stripes have nothing to do with defence against predators. Perhaps a herd of all-grey zebras would fare no better or worse where killers are concerned. The answer instead may lie within the world of zebras. Could it be that the stripes are intended not for the eyes of lions but for the eyes of other zebras?

One idea that supports this view sees the stripes as a means for the zebra to identify which species it belongs to. There are three main species surviving today. Up in the north of the range there is the big Grevy's zebra, with its very fine, narrow stripes. Over the whole of the middle of the range there is the common zebra, with many local races, all with big bold stripes. And down in the highlands of the south there is the extremely rare mountain zebra, with bold but more vertically arranged flank stripes. To human experts it is possible to identify each species at a glance, but to more casual observers a zebra is just a zebra, and this reflects the weakness of this argument. For if species identification was the function of the stripes they would have diverged much more in the three cases. If it was important for an individual animal to be able to distinguish between other zebras, evolution would have led the three patterns away from one another to a greater degree. In fact the opposite seems to have occurred because Grevy's zebra is in other ways a very different animal from the common zebra and it appears that in their stripes the two species have converged rather than diverged, over a long period of time. It must be admitted, however, that there is an exception to this in the rump region of the zebras' markings. There, each of the three species does have a very different and highly distinctive pattern of markings. In the Grevy's zebra there is a bold black central stripe surrounded by a bleached-out white area. In the common zebra the bold striping continues towards the central line. And in the mountain zebra there is an unusual, fine grid-iron pattern along the central line. In these rump displays the general zebra patterning diverges far more than on any other part of the body and it seems that in this restricted region, at least, species identification is operating. But this cannot, of course, explain the greater similarity in the striping of the rest of the body. The rump region, as is clear by looking at other hoofed animals, is the favoured place for species 'flags' of identity and there is a good reason for this. When animals are fleeing in panic they have the greatest need to 'follow the flag' of their own species, keeping the herd together.

One theory suggests that the stripes act as a species recognition device. The striped patterns of the three species of zebras shown opposite – common (top), Grevy's (middle) and mountain (bottom) – are not as varied as might be expected if they were important as species-isolating mechanisms. Seen from the side, they are at first glance very similar indeed. Only in the rump region do the stripes markedly differ from species to species. This means that, when fleeing at speed, each zebra can follow its own species 'flag'.

Another theory sees the stripes as a mechanism for personal, individual identification, rather like fingerprints. This could easily be used by other members of a herd as a way of 'naming' an individual zebra. A comparison of the markings on the faces of these four drinking zebras confirms how different each pattern is in fine detail.

And when they are fleeing it is, of course, the rump regions of the other animals that they see most clearly.

Another possibility is that the stripes operate at much closer quarters as a *personal* identification system. It is certainly true that no two zebras are exactly alike in the details of their black and white lines. It is as if each animal wears on its skin a giant fingerprint, with as much individuality as would be needed for personal recognition. It is entirely possible that this mechanism does operate, and that the members of a herd do use these clues when looking at one another. The weakness in the argument is that a herd of wild horses, totally lacking in the striped pattern, also has accurate individual identification and every member of every equine herd, including all domestic horses, is known individually to every other member. Without such abilities group organisation would be impossible, so stripes, although useful, are by no means essential for this. It is therefore hard to see personal identification as the main evolutionary pressure leading to the famous black and white pattern.

A more ingenious theory claims that the stripes operate as a form of visual bonding between members of each herd. They are thought to make each zebra feel it belongs more strongly to its group than it would do if its colours were nondescript or dull. The clear, vivid patterns excite the zebra's eyes and give it a sensation of strong identity, as if it were the member of a football team or a sports club with conspicuous striped shirts. Perhaps so, but if this is the case, why should vertical stripes be selected as the particular zebra pattern? This is where the theory becomes ingenious. It is argued that zebras, unlike antelopes, spend a great deal of time grooming one another's bodies, especially in the neck region. It is a friendly act of mutual benefit to the two animals involved, since they cannot groom themselves in this region. When they are engaged in this 'bonding' act, that helps to cement their friendships, their necks twist towards one another and this causes vertical folds or furrows to be formed in their flesh. Because of the way that the two animals stand in relation to one another while grooming is going on, it is these vertical lines that are placed slap in front of their 'nearside' eyes. All the while, as they groom, they see a vertical pattern of skin-folds. This striping becomes synonymous with social attachment and is therefore the ideal pattern for evolution to exaggerate with contrasting colours. Once established, the pattern could spread all over the body, until it reached the condition we know today.

This theory could explain the difference between zebras and the plain-coloured antelopes that live alongside them, but it fails to give an answer to the obvious question of why wild asses and horses lack stripes. The arguments that the desert-living wild asses in northern Africa live in such small groups that this type of bonding enhancement would not be needed, and that the northern horses have winter coats that are too shaggy for vivid striping, are somehow unconvincing. There were undoubtedly much larger herds of wild asses in the past and as for black and white markings being impossible on shaggy, cold-country coats it is only necessary to look at the shaggy-coated skunks and pandas to see that this is really not a problem.

Another, equally ingenious, theory sees the black and white coats as a display, not for predators or for fellow zebras but instead for insect pests. It is argued that disease-carrying insects and other small pests are reluctant to land on such a strange surface, with its intensely white and intensely black patches. The optical effect disturbs them and they flit off to land on some less vivid pattern elsewhere. In support of this theory is the fact that zebras are much less likely to suffer from certain insect-carried diseases than other equines. Ordinary domestic horses always suffered greatly in

A recent explanation is that the stripes are a mechanism for visual bonding. The stimulating pattern is thought to create a feeling of herd membership. In other words, the stripes function in much the same way as the patterns on football shirts in human sports groups.

A reason why a pattern of vertical stripes might have been selected as a social bonding device is that attachment takes place most strongly during periods of grooming (right), and this involves a zebra bending its neck, creating vertical folds in the skin (below); the bold stripes could have evolved to augment and elaborate these dark lines. Once established, the pattern would then simply have spread to cover the whole animal.

tropical Africa, and it could simply be that they attract more insects than zebras. But against this theory it has been argued that over large parts of the zebra's range, insects are not a major problem, so without further analysis the idea is once again inconclusive.

Finally, there is the even more unexpected theory that the black and white striping is essentially a cooling device. Everyone knows that white clothing is better in the hot sun than black, because the black allows more heat to be absorbed. It is claimed that by having sharply separated black and white regions of the skin, rotary breezes will be created between the black and white zones and that these will have a cooling effect. It is pointed out that the extinct quagga, a zebra with much less striping, came from the far south of Africa where the climate is cooler and that the northern, stripeless horses also enjoy cooler weather. In other words, the hotter the climate the more striped the equines become.

Unfortunately this theory does not stand up too well either, because the mountain zebra comes from the cooler south and yet has vivid stripes. Also, the wild asses from northern Africa survive in some of the hottest places on earth and their bodies are stubbornly stripeless.

Briefly, these are nine theories to explain why zebras have stripes. Each has something to commend it and yet each in turn is not totally convincing. Nothing could demonstrate more elegantly the fascination there is in studying animals today. The animal world is full of puzzles and half-solved mysteries of this kind. Every so often a new observation is made, a new fact discovered and another theory is formulated. Sometimes it is completely convincing and all the old ideas can be thrown away. More often it turns out to have weaknesses of its own and many a general theory ends up as tomorrow's 'special case', explaining something but not everything.

In the case of the zebra, it is entirely possible that several of the nine mechanisms are working together, each supporting the other. Or it may be that one of them alone is, after all, the true explanation of the vivid coat of stripes. Alternatively and more excitingly, perhaps around the corner there lurks a tenth explanation – the true one at last. We shall see.

On the pages that follow I have tried to present the most convincing and widely accepted explanations of all the various animal phenomena that are described there, but while reading them it is always healthy to keep an open mind. Few cases are quite as confusing as the zebra's stripes, but there is always the chance that in a year or so's time new information will have been gathered that changes our views about some aspect of animal life. If this were not the case it would be a dead subject, beautiful perhaps but boring. And the truth that I insist I *have* discovered about the animal world is that it is never, ever boring.

Grouping Behaviour

AMONG THE HIGHER forms of life, each species has a typical group size at which it functions most efficiently. Some animals are solitary, living on personally defended territories where no other member of their species is permitted to enter. They only relinquish this extreme form of isolation for a brief coming together of the sexes during the breeding season and then return to their separate lives once more.

Other animals pair for life, but never congregate into larger groupings. A pair of golden eagles may defend a territory of up to 16 square miles, living there in splendid detachment from the remainder of their species. In stark contrast, the red-billed quelea – a small African weaver-bird – lives in dense roosts of up to ten million individuals. They scatter during the day to feed, but reassemble every night in a vast, noisy crowd, blanketing whole sections of woodland and making the trees almost invisible under their cloak of bodies.

Even these 'feathered locusts' are outshone by the real thing. A typical locust swarm can easily contain 40 thousand million individuals and the record swarm – one that was said to stretch for 2,000 miles – was estimated at about 250 thousand million insects. An average-sized locust swarm devours in the region of 20,000 tons of vegetation every day. But these are not organised, structured animal societies – they are abnormal explosions of the locust population and they carry with them the seeds of their own downfall, as do modern human populations that are heading in the same direction. For we, like all animal species, have an optimum group size and it is one that we have exceeded so dramatically that our species is already well on its way to massive self-destruction. Like locust swarms we will experience a vast population crash at some point, one that will drag us back to a more natural level. But that is another story. In the meantime, what are the natural levels of social grouping for other animal species?

All higher forms of life live in groups of a particular size. Some are solitary, while others appear in huge flocks, like these flamingos.

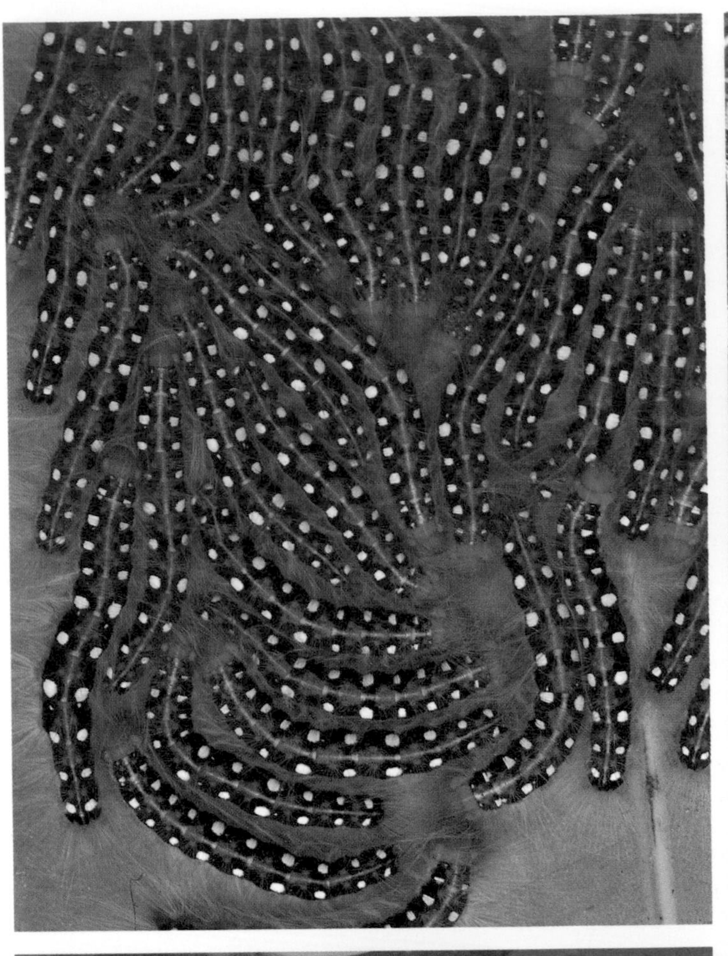

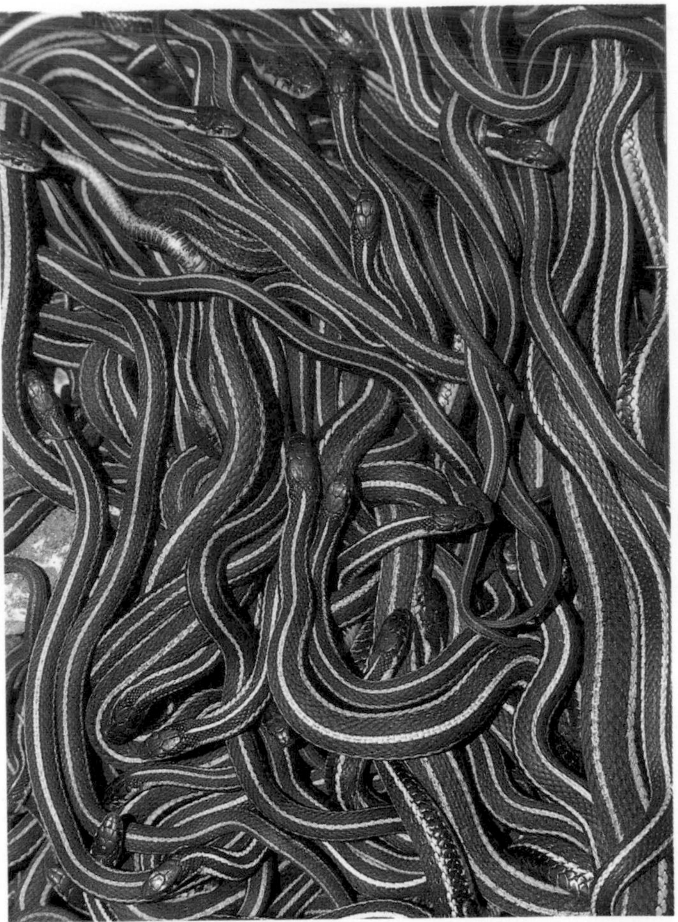

According to Noah, all animals go in twos, like the golden eagles, but in reality there are several basic social units, of which the isolated pair is only one example. To classify these units into different types is inevitably an over-simplification, but it is nevertheless a helpful exercise.

1 Solitary

The adult males and females live alone and meet only in the breeding season. After courtship and copulation they split up again and have no further relations until the next season. The female deals with the young without the help of the male, or in certain cases the male rears the young without the help of the female.

The solitary adults may wander over vaguely defined home ranges, or they may restrict themselves to clearly defined and defended territories. Animals that fall into this general category include many territorial fish, such as the river bullhead, many reptiles, especially snakes, a few birds, such as the ruff, and many mammals, such as bears, pandas and raccoons.

With birds and mammals it is nearly always the female that looks after the young, without the help of the male, but many fish reverse this procedure, with all parental duties being carried out by the males, as in the case of nest-building sticklebacks and pouch-carrying seahorses.

2 Pair

Wherever the parental duties are too much for one parent, the male and female form a strong attachment for one another and live together as a bonded pair. Nesting fish, like many cichlids, form pairs that share the parental duties equally, and the same is true of over 90 per cent of all bird species, where incubation and feeding duties are particularly demanding. Pair-bonds are comparatively rare in reptiles, and although they are

The red-billed weaver bird, or quelea, lives in colonies numbered in the millions (right). Each million birds destroys 60 tons of grain every day, making this species a major pest of African agriculture, rivalled only by the locust (above).

(Opposite) Impressive aggregations can be observed when caterpillars struggle for space in their breeding swarms, snakes congregate underground in hibernating masses, sharing their slight body heat, fish gather in vast, protective shoals and butterflies gather in large numbers on their wintering grounds.

Many animals, such as the puma, are solitary for most of the year, only coming together with other members of their species at mating time. For the rest of the year they wander their home ranges or defend their territories against all-comers.

Whenever parental duties require the attentions of both mother and father, pair-formation takes place (opposite, top). When a young female black-backed jackal first comes into heat she is followed by several young males, one of which she selects as her long-term partner. Later when she is denbound with cubs, he will bring food to her and the pair will then rear their cubs together.

Once the young are born they may be cared for by both their parents for many days. During this period, the family unit becomes the natural social grouping, as it is for these highly protective swans (opposite, bottom). A prolonged period of family life permits the growing offspring to add individual learning experiences to their inborn behaviour programming.

typical of certain mammals, such as beavers, jackals, foxes and gibbons, this is not the typical mammalian grouping. Female mammals are better at caring for their young without male assistance and, where it is necessary, there tends to be a larger group than just a single pair.

3 Family

This is merely an extension of the pair. When the young arrive they often remain with the parents until they are fully, or nearly fully, grown. During the later stages of their development to adulthood they frequently move about with their parents and, at this stage, create a larger grouping. Occasionally there may be overlapping litters, with older offspring remaining while new babies are being cared for. In such cases the extended family begins to look quite complex, but it is still essentially based upon the original pair, and when young do become fully adult they are driven away or wander off on their own to form separate pairs.

4 Harem

A typical mammalian grouping consists of one dominant male and a harem of females. The size of the harem varies from species to species. In polygamous monkeys each male has only a few females, but with species such as fur seals, the harems can be huge – up to one hundred females. Male deer and some antelopes are also capable of collecting and controlling sizable numbers of females in the breeding season, but whenever the harem becomes too massive the harem-masters lose control and small splinter-groups of females are stolen by less dominant males in the vicinity. The harem system requires, of course, that a large number of males are unlucky in the breeding season and must remain on the fringe of the reproductive groups. There they may become solitary or may form into bachelor groups. When the harem-masters become too old or too sick, new overlords will emerge from the strongest in the bachelor groups and will take over the females. Inherent in this type of social system is a marked sexual dimorphism, with males usually much bigger than females, a consequence of the fierce inter-male competition for dominance.

Apart from loss of control over females, there is also another danger in overinflating the size of the harem. In some polygamous species, such as the patas monkey, the males become henpecked. In this species, harems of up to twelve females are usual and although the male remains the leader, watchdog and defender of the group, in clashes with his females he is often the loser. There are enough of them to gang up on him and dominate him, even though he remains the group 'administrator'.

Some harems are permanent, while others dissolve at the end of the breeding season. When this happens, the individuals may disperse or they may stay together in new groupings. Red deer, for instance, end the season by splitting up into male herds and female herds, each with dominant members. Then, when the next breeding time comes around, the males start moving in on the female groups again and fight one another for possession of that year's harems.

5 Matriarchy

By a small shift in the balance of power between the genders, the harem system can be converted into a matriarchy. In this type of grouping, the females stay together at the centre of society, with the males on the periphery. Instead of the dominant males moving in and taking over the females during the breeding period, they are simply allowed in for mating only and then driven out again. This system is found in species as different as elephants and coatis.

When Victorian naturalists first studied the African elephant in the field they were convinced that the dominant member of the herd must be the

In many species the basic unit is the harem, with one dominant male surrounding himself with a number of breeding females. This is true of Hamadryas baboons (above), fur seals (left) and red deer (opposite, top). The number of females varies from species to species. The fur seal may have up to a hundred, but the Hamadryas makes do with only a handful. Harem-masters are always conspicuously different in appearance from their females – frequently they are larger and in many cases they have special male weapons on their heads.

As an outcome of the harem system there are many surplus males. In most harem-forming species these outcasts cluster together in all-male groups, as here in the case of young male red deer (opposite, bottom). From their number a new harem-master will one day emerge.

oldest bull. This fitted their human concept of the dominant father-figure. But we now know that the dominant leader of the herd is the old matriarch and that the big bulls are forced to live out their lives in solitary splendour on the fringes of elephant society. Occasionally several males may form loose groups together, but whether they do this or not the essential point here is that it is the combined efforts of the females that control the herd, and rear and defend the young.

6 Oligarchy

Moving up the scale towards more complex social groupings we come to the oligarchy, where power is invested in an élite gang of dominant males. The common baboon belongs to this type of social gathering. Each troop, as it moves about its home range, consists of several powerful males, their shared females and their combined offspring. Young adult males are driven out of this group and form separate bachelor parties that must bide their time until they can steal young females and set up an oligarchy of their own.

The advantage of this arrangement is that it provides an efficient means of defence against predators. When attacked by a leopard, for instance, all the dominant males rush towards the attacker and together are able to intimidate it. Singly, the leopard would defeat them, but together they can drive it away. They can also help one another to defeat an attack by rival male baboons that are attempting to take over their shared females. Furthermore, if one of the females becomes too dominant they can pool their male resources to put her in her place.

Some species have a female-dominated society. This probably evolved from a harem system where the females combined forces to overpower and drive out the dominant male. This matriarchal system operates in the African elephant where it is the dominant cow that leads the herd. The herd is made up purely of females and young.

This arrangement requires a certain amount of restraint and co-operation on the part of the dominant males, but it clearly has compensatory advantages. It is employed by a wide range of species, with slight variations. In some cases there is a strong peck-order operating within the group of males, and sometimes also within the group of females. With certain carnivores, only the dominant male copulates with the females. He never gives the other males a chance. In other groups, the dominant male gets most of the matings, and the other males are allowed an occasional copulation. Sometimes the dominant female is so powerful that only she becomes pregnant and her subordinates must await her demise before being able to breed themselves.

In wolves there is occasionally a drawback to being the 'top-dog'. If the pack has become too big and unmanageable, the dominant male must spend all his time trying to control it. He is often so busy that the less powerful males are able to disappear into a corner and copulate with his favourite females while his back is turned. This 'tycoon-failing' pattern is one of the factors that makes a more moderate pack-size a better prospect for power-hungry males.

7 Arena

Some species have developed an all-male grouping of a special kind. In various birds species, such as ruffs and black grouse, the males all cluster together in the breeding season on a special patch of ground called a lek or arena. There each displays as vividly as he can. The females visit the arena and select the male of their choice, mate with him and then depart to rear the young on their own. This is dealt with in more detail in the chapter on arena displays.

8 Hierarchy

A social hierarchy, peck-order, or social dominance grouping consists of individuals that are ranked according to their status. Whereas territorial individuals are dominant in their own area but subordinates in other areas, hierarchically grouped individuals have the same status wherever they go. Their rank is determined not by place but by person. Some individuals manage to intimidate all their companions and rise to the top of the peck-order. Others sink to the bottom. This arrangement can become part of a reproductive grouping, but it is also frequently observed in non-breeding groups, where a hierarchy develops in relation to food sources. The top animals can displace the bottom ones at feeding sites and gain other environmental advantages by virtue of being tougher than their group companions.

9 Aggregation

In non-breeding groups the size of the gathering can swell and swell, especially during migrations of the type seen in wildebeest and salmon. There, when animals are on the move, there is little social structure, merely a mass of advancing animals. The huge conglomerates that assemble on such occasions provide us with some of the greatest spectacles in the animal world. Vast swarms of insects, endless shoals of fish, huge herds of antelopes and sky-filling migrations of birds, all reach dramatic proportions and defy accurate counting.

10 Caste system

Among social insects there exist extremely complex social organisations in which there are different classes of individuals, providing an efficient division of labour. Termites, ants, bees and wasps display various forms of this caste system, all of which show a level of social differentiation that

In the common baboon, unlike the Hamadryas, the troop exists as an oligarchy, with several dominant males sharing power. This system is efficient because the males can gang up on attacking predators and their joint efforts are enough to drive them off. There is some degree of division of labour, with certain males acting as sentinels.

cannot be found among any vertebrate species other than man himself. There are queens, kings, workers, soldiers, and drones, each with a separate task to perform. Together they create a large nest, operate it smoothly and defend it against all enemies.

These are the ten basic types of social grouping. For the animal-watcher the fascination is in finding new variants of these themes. So often, a species appears at first glance to be a standard version of one of these ten categories. Then, on closer inspection, it emerges that there are oddities in this particular instance (as there are in so many cases) and the quest is on to identify the survival value of these particular idiosyncrasies. Usually it has something to do with the needs created by sharing: sharing the task of killing a large prey, sharing the task of defending the group against predators, sharing body heat to keep warm at night; sharing out different duties that can be performed at the same time; sharing the food that has been obtained by special members of the group; and so on. Each environment and each species throws up its own unique combination of social demands which, in turn, create the spectrum of variations on the theme of social grouping. And although we now know the facts for many species of animals, it is important to point out that for many more we still have only the crudest knowledge of their grouping behaviour. Much remains to be discovered.

Among the social insects dense aggregations occur but without the loss of complex social organisation. Bees, wasps, ants and termites exhibit extraordinary degrees of division of labour, with separate castes comprising such categories as kings, queens, workers, slaves, drones and soldiers.

Escape Behaviour

Most animals must live out their lives with an ever-present threat of attack from predators. When active, they must always be on the alert, ready to take avoiding action. They have three basic strategies: to freeze, to flee and to fight, often in that order.

Freezing is a widespread response to a predator alarm among many of the better camouflaged species. The eyes of predators are finely tuned to the slightest movement, but are less well equipped to distinguish static shapes, especially if their outlines are blurred by broken patterning or by foliage cover. Some species, such as hares and young antelope, will stay stock-still in a crouching posture until the predator is almost upon them, before bolting. Anyone who has been walking across grassland and has nearly trodden on a hare, will know what a shock it is when the animal suddenly leaps up and darts off. It is gone even before the predator can gather its wits and make chase.

Some prey species, such as tree squirrels, add a refinement to their freezing behaviour. As soon as they sight a predator approaching, they swiftly dart round to the far side of a tree-trunk before performing the rigid 'statue' response. Freezing on the trunk, they are now completely hidden from view and make no tell-tale sound. If the killer comes prowling around to their side of the tree, they simply dart to the blind side again. In this way, squirrels in a wood can remain hidden from view while a predator explores the whole region. Woodpeckers use a similar hiding technique. For the animal-watcher this strategy only becomes obvious when, from a hidden viewpoint, it is possible to see the predator approaching from the far side of the prey. Then the trick is revealed all too clearly.

If freezing fails to work, then the next step is to flee. Many prey species have developed amazingly fast escape actions. The speed of some of these have been carefully measured. Some deer have managed bursts of 40 miles an hour and antelopes have reliably reached over 50 miles an hour. The fastest predator is the cheetah, which can touch nearly 60 miles an hour at its best, but it cannot sustain this for very long. In fact, the average distance covered during pursuits by cheetahs is only 183 yards. In exceptional cases they may be able to continue for 300 yards, but beyond this they quickly become exhausted and give up. This is where the prey can score, because they can keep going for much longer. Some antelopes have

When under attack from a predator the three basic responses of animals are to freeze, flee or fight. In the bittern (above) and the hare (left), freezing is the most common form of defence. At the first sign of danger the bittern adopts a rigid, stiffly erect posture, making itself almost invisible in the vegetation around its nest. The hare crouches low on the ground and remains absolutely still until the predator is nearly upon it. Only then does it flee. Even without elaborate camouflage markings, this immobility is of great value to prey species because the eyes of prowling killers are more sensitive to movements than to static shapes.

Some prey species, such as the oryx (opposite, top), rely on simple, powerful galloping to escape their pursuers, but others, such as the impala (below), employ a more confusing, zigzag escape route, with sudden directional changes making it difficult for the pursuing attacker to strike.

(Opposite, bottom) Many prey species have developed unusually long legs and strange forms of locomotion as refinements of their fleeing behaviour. Bipedal leaping in kangaroos and bipedal sprinting in frilled lizards help to make their flight more startling and unpredictable.

been recorded fleeing at 42 miles an hour for a mile, or 35 miles an hour for 4 miles, and deer have been able to sustain 30 miles an hour for an astonishing 20 miles. This may seem excessive in relation to the cheetah's performance, but there are other predators with much greater stamina. Wolves and hunting dogs may not be able to reach dramatically high speeds over short distances, but they can run and run. The speed of one wolf was recorded at between 15 and 30 miles an hour for a distance of 12 miles, after which the animal had to slow down to a trot. Tundra wolves have been observed chasing deer for 5 or 6 miles before accelerating to attack. Clearly prey need to be both sprinters over short distances and marathon runners over longer ones, to be able to defend themselves against both types of pursuers. The secret against sprinting predators, such as the big cats, is to get a good start on them. As they are marginally faster, the ability to begin fleeing before they have had a chance to accelerate to full speed is crucial. Hence the ever-alert nature of so many antelopes. They must always be on the lookout for the stalking feline hunter, whose aim is to get as close as possible before beginning the chase. Once the pursuit is on, it is simply a matter of whether the killer will start to tire before it manages to reach its victim. With wild dogs in pursuit it is a different matter. Then it is a battle of relative staminas – who will last the longest.

If attacks come from above, running fast is less effective. Birds of prey have been registered nose-diving at speeds of up to 110 miles an hour. The best defence here is to go underground as quickly as possible. The main prey of predatory birds are the hundreds of different rodent species and the vast majority of these have evolved a protection system based on a quick scamper down a tunnel. The burrow has become their main survival device, saving them not only from swooping birds but also from earthbound predators too large to follow them.

Many other prey species flee upwards to safety rather than downwards. Arboreal rodents, primates, and many other mammals take to the trees as an escape route and the majority of them rarely risk a descent to the

The hare, pursued here by a goshawk, is a master of unexpected movements when fleeing.

Chasing prey through the tree-tops is much more hazardous than pursuing them along the ground, as this raccoon-hunting puma is discovering (below). Many species have taken to arboreal living as a protective device, and some, such as this sugar-glider (opposite), have become specialised jumpers that can leave most predators behind with ease.

ground unless it is vitally necessary. Birds, bats and gliding mammals simply take to the air. Many of the most impressive forms of animal locomotion have evolved as extensions of fleeing actions.

There are several refinements of fleeing behaviour. Instead of trying to escape by moving as fast as possible away from the predator, some animals take an erratic zigzag course. The hare flees across a field in this way, and a number of different birds and fish have also been seen to use this technique. It works on the principle that the pursuer will not be able to change direction as efficiently as the prey. To be successful the direction-shifts of the fleeing animal must be irregular so that the predator cannot anticipate either when or in which direction the next change of course will be. Unpredictable escape patterns of this kind have been called 'protean' defence systems because like the figure of Proteus in ancient Greek mythology they repeatedly change their shape.

Another technique is the dash-and-hide, dash-and-hide retreat. This is employed by animals such as frogs and grasshoppers that cannot sustain a prolonged bout of fleeing. They make a sudden darting movement at the highest possible speed and then quickly freeze, staying quite still in the undergrowth. If the predator takes a wrong direction they remain in hiding until the danger is passed, but if it comes close again they wait until the last moment and then repeat the dash-and-hide movement. They may keep this up time and time again until, with luck, the predator finally gives up the chase.

A third device is the flash retreat. In this the prey flashes a bright patch

(Opposite, top) Self-defence for a rhinoceros may be little more than casting a sleepily arrogant eye on the hungry lions surrounding it; for an agitated elephant, on the other hand, it may be a vigorous charge.

Horned prey, such as these musk oxen (opposite, bottom), are only armoured at their front end, so that their rear needs special protection. The herd solves this problem by forming a defensive circle when marauding wolves attack.

of colour as it flees, and then it suddenly freezes, concealing the bright patch as it does so. Some grasshoppers have vividly coloured patches on their wings and these are clearly visible when they are flying away from their enemies. But as soon as they come to rest the wings close and the animals become completely cryptic again. This confuses the predator because it has 'tuned in', so to speak, to the bright colour and is following that. Once the colour disappears it is as if the prey animal has suddenly ceased to exist.

If fleeing and hiding have failed and the prey finds itself cornered by its would-be killer, it may make one last desperate attempt to save itself, by throwing caution to the wind and attacking the predator physically. In most cases this will not be of much use, but at this stage there is nothing to lose. And predators are surprisingly nervous about sustaining even minor injuries. This is because, should they be wounded, or should a small cut become infected, they may temporarily lose peak condition. If they are not athletically fit they may not be able to hunt for a while and this could be critical, as most of them must kill regularly to survive. So a snarling, spitting, clawing prey is given rather more respect than it may appear to deserve. Many a hungry cat has thought twice about attacking a large, cornered rat. Many small birds have escaped with the loss of only a few feathers because of a smartly delivered peck with a sharp beak. And animals with horns have sometimes managed to toss their tormentors high in the air when making a last-ditch stand. For some prey, it becomes a matter of safety in numbers, with musk oxen forming up in a defensive circle to face a pack of marauding wolves, like a wagon train defending itself against attacking Indians.

With these three techniques of freezing, fleeing and fighting many prey manage to survive for another day, but there are also more specialised ways of dealing effectively with attackers and these are discussed in more detail in the following chapters.

If fleeing fails and the prey is cornered, it may make a last-ditch stand and turn on its tormentor. Self-defence of this kind may be no more than a threat, as in the case of the inflated toad, standing high on its legs when confronted by a grass snake (above), or the frilled lizard (right) which erects a huge umbrella of skin around its gaping jaws.

Protective Armour

THE ANIMAL WORLD is so full of savage teeth, sharp claws and pointed horns that many species have sought protection behind a powerful wall of body armour. Once they possess this their whole lifestyle is transformed. Gone is the need for speedy, athletic fleeing, gone the need for eternal alertness and high-strung sensitivity to every tiny hint of danger. They become comfortably heavy-bodied, even cumbersome, and bumble their way through life at a more relaxed pace. They are the tanks of the animal kingdom, and they come in many forms.

Among the mammals they are comparatively rare. Only 169 mammalian species (which is roughly 4 per cent of the 4,237 living species) can boast any form of armoured protection. It is true that there are many species with horns or antlers, but those are primarily concerned with offence and are only secondarily employed as defensive armour against predators. They are weapons rather than protective coverings.

The most primitive of the armoured mammals are the five species of echidnas, or spiny anteaters, from Australasia. Their underside is soft and hairy but the whole of their upper surface is covered in strong, thick spines. When disturbed, they make for a crevice or hole where they can jam themselves in so tight with their powerful claws that it is almost impossible to dislodge them. All the attacker is offered is a sea of spines, with nothing to grip on to. If caught out in the open, the echidna simply digs rapidly downwards until, again, nothing is visible but its spiky upper surface.

Among the insectivores there are eight species of spiny tenrecs from Madagascar and fifteen species of hedgehogs from Europe, Africa and Asia. The tenrecs look very much like the hedgehogs and have a similar defensive armour, with fine spines that can be erected as the animal rolls into a tight ball. Some of the hedgehogs embellish their self-protection with an upward jerking movement. If a predator is nosing close to them and they have not yet rolled tightly into a ball, they may bend and then suddenly straighten their legs, thrusting their spiny backs up into the

The echidna, Australia's spiny anteater, protects itself by digging rapidly downwards until its soft underside is completely hidden and its enemy is offered only a sea of powerful, sharp spikes.

The hedgehog cannot dig itself in when threatened but it can protect its soft parts by rolling up into a tight ball. Its seven thousand spines then defend it in all directions. The spines also act as shock-absorbers if the animal suffers a fall.

killer's sensitive nose. This upward jerk can be very painful and is sometimes enough to dishearten the attacker and allow the hedgehog to scuttle away to safety without having to resort to the passive rolling-up defence.

In addition to these spiny insectivores there are two other armoured species, the extraordinary hero shrews. Outwardly these look like large, furry shrews, but internally they carry a secret defence – their amazing, reinforced backbone. Unlike the spinal column of any other mammal, it has a network of interlocking protrusions from its vertebrae that gives it enormous strength and resistance to crushing. Confined to West Africa, these species of shrews have been persecuted locally because of their strange internal armour. Local natives believe that, eaten or worn as decoration, they will act as protection from injury. To demonstrate the power of the animals, a fully grown man stands on the back of the 6-inch-long shrew, balancing himself on one leg. Any other small mammal of this size would quickly be squashed to death, but when he steps off it after several minutes, the hero shrew reveals that it is still very much alive by trying to escape. The mystery about this particular form of protective armour is precisely what the shrew is defending itself against. Humans were clearly not the primeval enemy that led to the evolution of this unique backbone, so which deadly crusher was involved? This is a puzzle that has yet to be answered.

Another mystery surrounds the only example of body armour to be found among the primates. The more advanced monkeys or apes do not have any kind of protective armour and use brain rather than brawn to solve their problems, but the slow-moving little relative of the bushbabies known as Bosman's potto does have a secret defence rather like the hero shrew's. Here, too, the backbone is modified and made more impressive, but the details are different. In the potto's case, the neck vertebrae have

The hero shrew has an armoured backbone that can withstand enormous pressure.

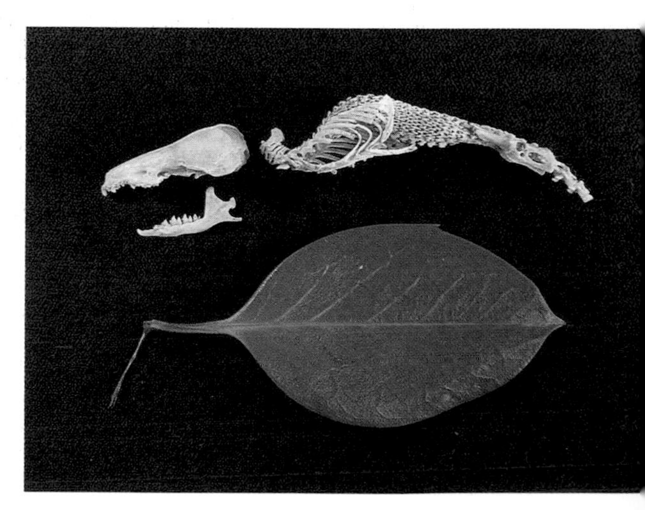

This spiny mouse is one of 101 species of prickly rodents, but it is the least spectacular. At first glimpse it may not even appear to be armoured, but if the animal is taken into a predator's mouth its fur is sufficiently spiky to cause acute discomfort and it is quickly dropped.

short, blunt vertical protrusions that project slightly through the skin. The result is a kind of knuckle-duster, a row of hard bumps hidden deceptively in the soft, dense fur. The puzzle here is how the animal employs these obviously defensive spikes when in trouble. When the slow-moving potto is approached by an enemy, it curls up into a tight ball, clings on to its branch and lowers its head between its legs so that the back of its spiky neck is thrust towards the predator. But this hardly seems sufficient, as an armoured defence, to deter any but the most casual of killers. Perhaps, as the attacker's jaws open wide to grab the small, furry shape, the potto acts like a jerking hedgehog and thrusts the back of its neck at the biting mouth. Possibly the shock of feeling hard spikes in what appeared to be a soft, yielding morsel of food would be enough to make some predators think twice, but this is not particularly convincing. If the spikes were sharp, like the dorsal fins of many fish, the strategy would be easy to understand, but their shortness and bluntness mean that, for the moment, the potto remains another puzzle.

Sharper spines, including the most fiendish known, are to be found among the rodents. In this huge group (1,729 species) there are many armoured examples: 18 kinds of spiny mice, 22 spiny pocket mice, 37 spiny rats and 24 porcupines. The mice and rats have simple, slender spines mixed in with their ordinary fur. Although they are not completely spiny, they are nevertheless prickly enough to repel all but the most desperate of predators. It is with the porcupines, however, that the spiny defensive system reaches its zenith. Here the spikes can become enormous quills – those of the great crested porcupine are as long as 20 inches. They can be rattled to produce a menacing warning sound, they can be thrust violently into the flesh of the attacker by backward lunges of the porcupine's body, and they can easily be detached to be left embedded in the unfortunate predator's anatomy. Once driven home, these savage spines are extremely difficult to dislodge and may continue to torment the would-be killer for days to come, sometimes leading to festering wounds that can kill.

Less spectacular but even more deadly are the spines of the American

The most dramatic of the spiny rodents are the heavily armoured porcupines. The South American porcupines (below, left) may not look as impressive as the great crested species from Africa (below, right) but their quills are sharply barbed so that once they have entered a predator's flesh they cannot easily be removed.

porcupines. These spines are covered in small backward-directed barbs, so that once in the flesh they are almost impossible to dislodge. Furthermore, the pain caused by the presence of an embedded spine makes the muscles it has penetrated contract in such a way that it is driven even deeper into the flesh. Some barbed quills have reputedly travelled so deep into the bodies of their victims that they have caused the deaths of the predators from the extensive damage caused to vital internal organs. It has been calculated that the $1\frac{1}{2}$-inch quills penetrate tissue in this way at a rate of an inch a day. The North American porcupine's method of defending itself is to lash wildly at the enemy with its spiny tail, sometimes turning the predator's nose into a pin-cushion. The only killer that seems to have much success in dealing with this troublesome prey is the North American marten known as the fisher. This species circles the porcupine like a mongoose dealing with a cobra, and confuses it enough to be able to strike at its only vulnerable zone – its head. After a few blows there, the porcupine succumbs and is then turned carefully over and eaten from underneath, its soft underbelly causing no problems. Eventually all that is left is the spiny coat.

Mammals with non-spiny armour include the twenty-one species of armadillos and the seven species of pangolins, or scaly anteaters. The upper skin of the armadillo's body has evolved into a hard horny covering that is hinged in such a way as to give the animals at least some degree of flexibility. The hinges are arranged in bands across the middle of the back and vary in number according to the species. When alarmed, these animals usually try to make for cover, seeking out their burrows or some other hiding place where they can wedge themselves in as tightly as possible. The smallest of the species, the pink fairy armadillo, possesses one of the most extraordinary rumps in the animal kingdom. Its rear end has become a completely flat, circular disc of armour. When the little animal is disturbed it burrows furiously down into the ground until it has completely disappeared except for its horny rump. This acts like a cork in a bottle, stopping the borrow and preventing any attack on the soft flesh beneath it.

The spines of the African porcupines regularly encountered by lions may not be barbed, like those of the South American porcupines, but they are nevertheless difficult to remove and this particular lion faces a painful future.

The most impressive of all the armadillo species, when responding to attack, is the three-banded. This animal rolls up into an almost perfect ball, with the top of the head and the top of the tail fitting close together like the last pieces of a jigsaw puzzle, leaving no gap anywhere for a predator to probe. As an added deterrent, this species usually waits until the last moment for the final clamping-down movement. It rolls up so that the gaps between its different parts are nearly closed, but a small opening is left. Then when it is touched it immediately snaps the shells together like a steel trap. If, at this point, a tender nose-tip, or small paw is being probed into the cracks, the result is both painful and startling, encouraging a more cautious approach to this difficult object of interest.

The armadillos are all confined to the Americas. In the Old World they have a counterpart in the form of the strange, scaly anteaters, the pangolins. Looking like huge, animated pine-cones, these unusual animals are covered in large, overlapping scales that extend right down to the tip of the long, prehensile tail. When disturbed they roll up so tightly that it is virtually impossible to unroll them. Although well protected against most predators they are no match for human ingenuity. Their flesh is popular with many local peoples and their scales are valuable as ornaments and decorations, with the result that their numbers have been dwindling.

Equally vulnerable to human weapons are the armoured giants of the animal world, the rhinos and the elephants. The most impressive of the rhinos is the great Indian rhinoceros, with its tough skin thrown into heavy folds that look like curved, welded sections of sheet-metal. Even the tail is carefully protected, fitting snugly into a deep groove on the animal's rump,

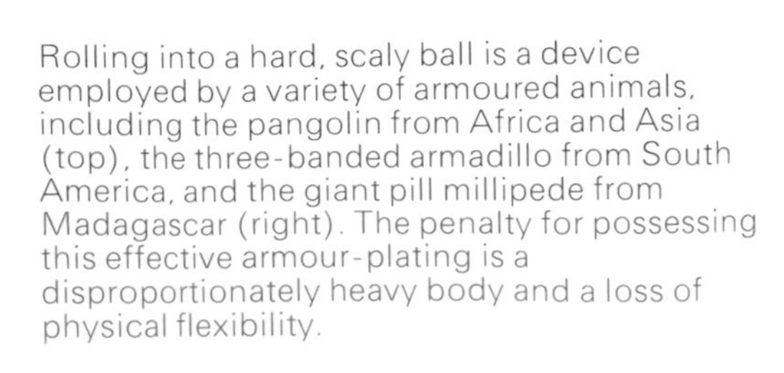

Rolling into a hard, scaly ball is a device employed by a variety of armoured animals, including the pangolin from Africa and Asia (top), the three-banded armadillo from South America, and the giant pill millipede from Madagascar (right). The penalty for possessing this effective armour-plating is a disproportionately heavy body and a loss of physical flexibility.

where two sheets of armour meet in a vertical cleft. Elephants lack these conspicuous folds, but they too are well defended against attack by their powerful, leathery skin which is sometimes as much as an inch thick.

Among the birds, body-armour is absent for the obvious reason that it would make them too heavy to fly. Only the large, ground-living cassowaries from the Australasian rain forests possess anything that could be called avian armour, and even there it is confined solely to the head region. The top of the skull is extended into a bony helmet, called a 'casque', that helps to cut a swathe through the dense undergrowth when the big bird is moving fast. It shields the head from damage and can therefore be described as protective armour of a limited kind.

Only the legs of birds show the tough scaly skin that reflects their reptilian ancestry. It is to the reptiles themselves one must turn to find the greatest display of armoured scales that are all-enveloping, protecting their wearers over the whole surface of the body. Each scale is a hard, dry shield of keratin growing out from the animal's epidermis and overlapping with its neighbours to make a body-covering that is as snugly fitting as it is strong. The keratin is dead material that is replaced by growth from below. The old armour is either worn away gradually, as in crocodiles, or it is shed at special times in a complete 'coat', as in lizards and snakes. The strongest forms of reptilian armour are reinforced by small plates of bone lying under the horny scales. These are at their most impressive on the backs of the giant crocodiles, where they are lozenge-shaped and give these

The great Indian rhinoceros carries a heavy burden in armoured skin. This is a handicap when racing a pursuing predator, but at the same time makes the rhinoceros an almost impossible meal. Even its tail is protected from attack, fitting neatly into a slot in its rump armour.

The scaly skin of reptiles is ideally suited for development into sharply spiked armour. The armoured scales become extended into thorny backs, horny heads and spiny frills. Some species, such as the armadillo lizard (above, left), make predatory attacks even more difficult by biting their tails and holding tight, creating an awkward circle of spikes. The thorny devil (above) relies simply on its immensely sharp and powerful spikes for protection, while the bearded lizard (left) defends itself by erecting its spiny throat skin.

formidable reptiles the appearance of being covered in spiked metal armour-plating.

Certain lizards take the spiky defence trend a major step further, with long, sharp spines growing out of their scaly skins. As painful to bite as cacti, spiny lizards such as the mountain devil or the horned toad are almost impossible to swallow without serious damage to the health of the predator.

Among the tortoises and turtles, the overlapping scales have been replaced by a large, thick armoured box of horn. In certain species this can be shut tight as a further protection, with the head and legs being drawn behind the barrier. Some have a hinge to close the head-end aperture; others have a hinge to protect the hind legs; still others have both. The

Body-armour is rare among birds but the ground-living cassowary (opposite) carries a powerful helmet on its head that acts as a protective shield when the bird pushes through rough undergrowth.

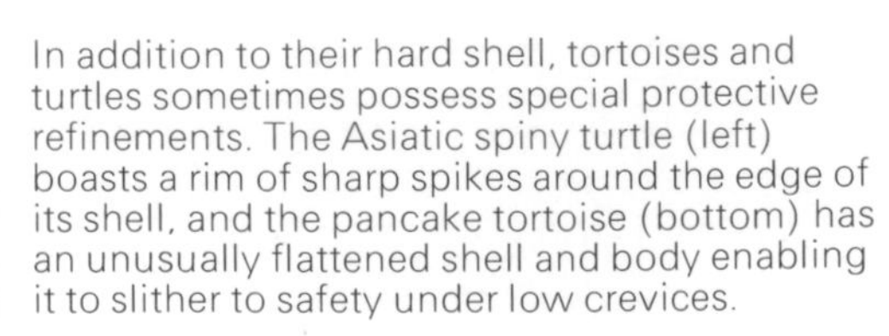

In addition to their hard shell, tortoises and turtles sometimes possess special protective refinements. The Asiatic spiny turtle (left) boasts a rim of sharp spikes around the edge of its shell, and the pancake tortoise (bottom) has an unusually flattened shell and body enabling it to slither to safety under low crevices.

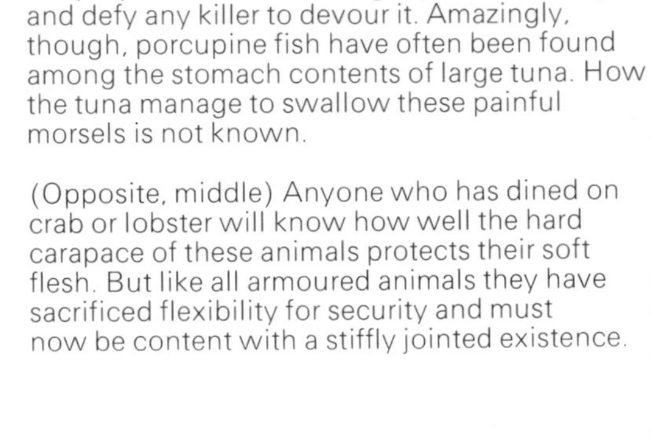

The porcupine fish (opposite, top) inflates itself by gulping in water (or air if it has been removed from the sea) until its body is so puffed up that its spiny scales stand out rigidly from its body and defy any killer to devour it. Amazingly, though, porcupine fish have often been found among the stomach contents of large tuna. How the tuna manage to swallow these painful morsels is not known.

(Opposite, middle) Anyone who has dined on crab or lobster will know how well the hard carapace of these animals protects their soft flesh. But like all armoured animals they have sacrificed flexibility for security and must now be content with a stiffly jointed existence.

bigheaded turtle is unique in having a heavily armoured head and tail which it never attempts to draw into its shell when attacked.

Amphibians, like birds and most mammals, have soft bodies that are easily squeezed in predatory jaws. They rely on many forms of protective behaviour, but the wearing of armour is not one of them. There is only one notable exception to this amphibian rule and that is the extraordinary sharp-ribbed newt. This small and seemingly harmless creature carries a hidden weapon. When it is grabbed, its elongated, sharp-pointed ribs are pressed out through the sides of its body and into the lining of the mouth of the hapless hunter. The tips of the spiny ribs pass through special pores as they impale the predator and they carry with them the secretions of nearby glands – glands that cause intensely painful poisoning of the inside of the attacker's mouth. Such newts are long remembered and long avoided.

Many species of fish employ similar defence tactics, with fin-rays hardened into sharp spines which are often greatly elongated and frequently have poison added, making a lasting impression. Some fish

The greatest armour in the world of molluscs is found on the giant clam (opposite), a bivalve whose shell weighs $\frac{1}{4}$ ton and is over 4 feet in length. Although this amazing armour protects the soft body of the clam extremely well, the animal is not the killer it is so often made out to be. Countless stories exist of swimmers whose feet become trapped inside giant clams that snap shut suddenly and then refuse to release their victims. This myth is believed all over the world despite the fact that the weight of the armour makes it impossible for the clam to close rapidly. There is no recorded example of a giant clam causing a human death.

boast armour so deadly that the predator dies within minutes of biting the spines. Observations of predatory fish attacking spiny prey reveal that after one experience of even a non-poisonous species the predators are so shocked by the pain inside their mouths that they go off their food for days. Eventually, when they begin cautiously to feed again, they avoid anything remotely like the spiny prey they attacked previously. And the bigger and sharper the spines, the longer this effect lasts. The supreme example of spiny defence is shown by the porcupine fish, a small inflatable creature that can puff itself up into a prickly sphere, either with an intake of water when under the surface, or with air when it has been caught and removed from the sea. As its body becomes fully inflated all its spines protrude menacingly, defying any predator to attack. At other times these spines lie flat against the body and are inconspicuous.

Spines are common among the lower forms of life and are clearly the most successful of all armoured devices. There are treacherous spiny sea urchins, spiny molluscs and spiny lobsters, all well protected by their sharp-pointed defence barriers. Even without spines, nearly all arthropods (prawns, crabs and lobsters) and molluscs (snails and bi-valves) are snugly enclosed in hard outer coverings. Their shells provide them with valuable protection but like all armoured species they are at certain disadvantages – they are nearly always less mobile, less flexible, and suffer the consequences. They are also less cunning and exploratory, having less danger to face than the soft-bodied species. In general, armoured animals are like rich people – insulated from the real hazards of life and therefore less

'streetwise' and opportunist. It is no accident that all the higher forms of life, all the most intelligent species, are without armour.

The bumbling nature of armoured animals is neatly summed up by an anecdote concerning a family of three-banded armadillos. They were trotting along a path on the side of a hill when they detected a man walking below them. When they immediately reacted with their automatic alarm response of rolling up into a tight ball, the entire family promptly rolled down the slope of the hill and came to rest at the man's feet, where he picked them up and popped them into his collecting bag. Sometimes it pays to be a little more flexible.

The long spines of the sea urchins protect them from most predators. However, this sea star, like many starfish, is capable of devouring these armoured animals. It can protrude from its body a soft, slimy stomach which then glides between the spines and invades the centre of the sea urchin. The victim's body is slowly digested before the stomach is withdrawn again to the interior of the sea star.

Camouflage

If an animal is to survive it must either confront predators or avoid them. If it possesses some sort of strength – physical, chemical or armoured – it can face up boldly to its enemies and defy them. If it is harmless, tasty, or soft-bodied, it must find some way of concealing itself. Under duress it may flee and hide, but it can only do this occasionally. For the rest of the time – the quieter moments in its life – it must render itself inconspicuous. It must conceal its form with some sort of camouflage.

The most obvious clue to its presence is its movement, so in order for its camouflage to work it must be prepared to remain very still for long periods of time, or at least to alter its position very slowly and gently. Its next problem is how to conceal its characteristic outline. Most animals are bi-laterally symmetrical and their shape is too balanced to match with the irregularities of backgrounds made up of plants, rocks and earth. To hide their clear outlines they need some kind of *disruptive markings*.

There are several types of disruption. One of them employs a psychological trick. If an animal is covered with irregular patches that break up its outline, the concealment can be improved by the presence of bright colours within these patches. These intense areas automatically draw the eyes of the predator and as he stares at them he fails to notice the larger shape which carries them. This is the Harlequin Principle and is vividly expressed on the body of the harlequin sweetlips, a fish that appears to the eye as a collection of coral reef growths.

Another type of disruption employs a series of intense bands or stripes that pass right across the animal. These are nearly always pale and dark in alternation and again help to chop up the animal's basic shape. If the animal is seen against a pale background, the pale patches blend in with the environment, leaving only the dark patches visible. On their own, these do not add up to an animal shape, and disruption is achieved. If the background changes to dark, the same system operates. This is the Particoloured Principle and is found in a wide variety of fish and insects.

A third type of disruption helps to conceal particular parts of the body. Certain anatomical shapes such as legs, arms, necks, feet or wings tend to give the game away. They help to define the prey's body *as* a body. A valuable way of concealing them is to have bars, stripes or blobs of colour that appear to overflow from one section to the next, as though someone has splashed ink across the animal. This Splash Principle is finely developed in many frogs. When they are moving about and their legs are held apart from the body the splash-pattern is not obvious, but once they crouch motionless, with their legs drawn up to their sides, their trunk and leg patches meet and obscure the divisions between them. Most disruption works by making continuous outlines appear to be discontinuous, but the Splash Principle operates in the opposite way, making what is in reality a discontinuous surface appear to be a continuous one. In both cases the camouflage is interfering with the natural elements, but in one case it is attacking the whole shape, while in the other it is blotting out the appendages.

More generalised disruptive markings which simply break up the broad shape of the body are found in the many spotted or striped species. Big cats such as the tiger employ stripes that work well in the grassy undergrowth. Serval cats, ocelots and leopards favour spotted coats that work well in dappled light. To human eyes many of the body markings appear to be beautiful and highly conspicuous, but this is because we so often see them in artificial environments where their camouflage quality fails to show itself. Seen against a plain, bleak background, most disruptive patterns fail miserably, but there are few such backdrops in nature.

One highly specialised form of disruptive pattern is the eye-mask. This takes the form of a stripe, band or patch of dark colour that runs right

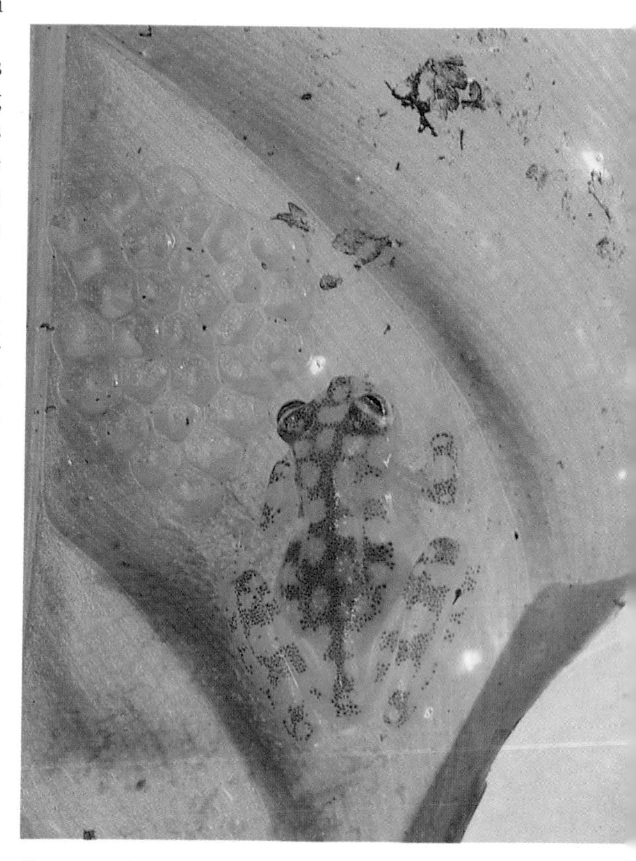

Spots, stripes and dappled markings help to disguise the presence of many animals. This small glass frog guarding its eggs on a large green leaf combines a semi-transparent body with a spotted skin that renders it almost invisible to its enemies.

The Splash Principle – the dark markings on this grey tree frog spread across its legs in such a way that they break up its shape. They look as though they have been splashed across the skin.

The spots of the leopard (opposite) lurking in the long grass help to destroy the solidity of its form, and the dappled patches on the coat of the clouded leopard (above, left) conceal it well in the broken light of the forest. Even the huge bulk of the tiger (above) is inconspicuous as the animal lies quietly on the forest floor.

through the eye and effectively makes it invisible. The eye is the most telltale detail of any animal shape and it is a great advantage to be able to obscure it from view. If the eye's owner employs the more obvious technique of closing the eye, it is no longer able to monitor the predator's movements. But if the eye can remain open without being seen, then the prey has the best of both worlds. This eye-mask device is extremely common among fish, with literally hundreds of species employing it, and it is also widely seen in snakes and frogs. A special variant of the eye-mask is the spot-the-eye pattern of certain fish. In this, the markings consist of a pattern of dark spots all of which look like the dark pupil of the fish's eye. In this crowd of spots, the real eye is lost.

A distinct form of concealment involves camouflage by *countershading*. This works on a completely different principle from disruptive colouring. Instead of trying to break up the real shape of the animal, it attempts to make it appear flatter than it is. All animals have three-dimensional bodies and their solidity can give them away. If they appear to be two-dimensional they suddenly seem less 'animal-like' and may avoid the unwelcome attentions of a hungry predator. Countershading operates on a simple premise, namely that in the animal's usual position one part of its body is more brightly lit than the rest. A completely plain, fawn-coloured antelope standing in a fawn-coloured landscape would appear to be solid-bodied because the sun, beating down on its back, would illuminate this area more than its sides and much more than its belly. In many species the natural coat colouring compensates for this top-lighting, the back being darker than the sides and the sides darker than the belly. In this way, the combination of light-plus-colour gives the animal a flat, undifferentiated appearance and renders it much less conspicuous to predatory eyes. The same principle is at work in some caterpillars, except that, since they are usually crawling along twigs and branches upside-down, their countershading is reversed. If they are deliberately turned up the 'right' way, they appear doubly solid, as the brighter light strikes the paler area.

The simplest method of all, where body concealment is concerned, is to

Disruptive markings help to break up the shape of an animal's body in any environment that is full of irregular detail, as in the case of this well concealed crab.

match perfectly the colour of the background. Whether this is a plain green, or yellow, or an overall mottled grey and brown, the result is a body-covering of precisely the same hue and tone. Tiny, pure green frogs live out their lives on large, pure green leaves. Nothing more elaborate is called for. Arctic mammals and birds have evolved pure white fur or feathers that help to camouflage them against the white expanses of the frozen north. Gravel-bottom-dwelling fish have gravel-pattern colours. But the problem often becomes more complex than this. As the animal moves about, its background changes. Dealing with this problem, many species have evolved various forms of colour change. Some, like flatfish and chameleons, are capable of changing colour rapidly by expanding and contracting special pigment cells in their skin. The octopus can even modify the texture of its skin at high speed, a smooth, plain surface suddenly furrowing up into complex folds and ridges and gaining a complex blotched patterning at the same time. Other species are only capable of undergoing such changes at a much slower rate – as when birds moult in and out of their breeding plumage. In the mating season finding a partner is more important than camouflage and bright colours are moulted in, but during the rest of the year self-protection dominates and camouflage returns.

The colours of certain other species can be modified only when there is a genetic shift in their population. This has happened recently with the peppered moth, a speckled species that is camouflaged on the lichen-covered tree-trunks where it normally lives. Following the Industrial Revolution, many trees became blackened by smoke and on these the speckled moths were suddenly highly conspicuous. They were an obvious target for hungry birds and were easily picked off. Only moths that were unusually dark managed to escape, and these then bred to produce a generally greyer strain of peppered moths. Again, it was the darkest ones that survived best, with the result that in a few generations the peppered

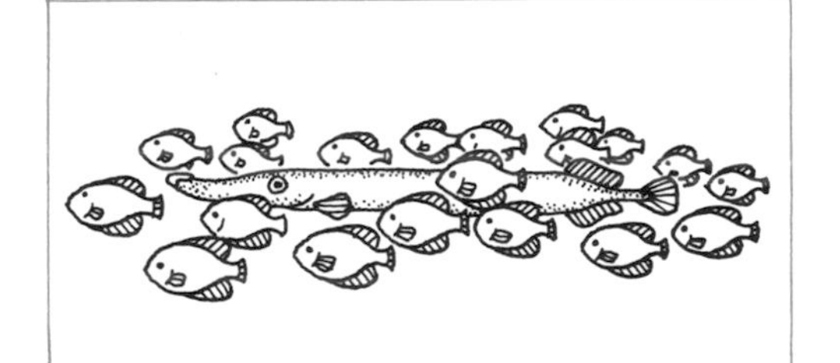

The trumpet fish is a lurking predator that often hides in shoals of smaller fish.

When natural light falls on an animal's body from above, it illuminates the upper surface more than the lower. As a result, many species, such as this predatory trumpet fish, are counter-shaded – their undersides are lighter than their backs. This compensates for the top-lighting, making the animals look flatter and less conspicuous. When artificially illuminated from below, the effect of the countershading is reversed – the trumpet fish looks strikingly solid and highly conspicuous.

moths living near the big industrial cities were nearly all black. By 1900 a survey near the city of Manchester revealed that, there, 99 per cent of all the peppered moths were already black, while those continuing to live in cleaner, country districts retain the usual, speckled appearance of the species. In a very short space of time – just a few decades – the industrial moths had gone through a small but distinct evolutionary step. Today the melanistic moths are still thriving in the grimier habitats and, if our cities go on spreading at their present rate, these black ones may well become the common form of the species.

Perhaps the most extraordinary examples of animal camouflage are those in which the animals concerned have evolved into mimics of specific details in their backgrounds. There are insects that look exactly like green leaves. Others are virtually indistinguishable from the flowers with which they associate – so much so that smaller insects keep settling on them. There are stick insects that look like twigs, butterflies that resemble bird droppings, seahorses that appear to be pieces of floating seaweed, moths that look like bark and frogs that seem to be nothing more than leaf litter on the forest floor. The tiny perfections of detail in these deceptions have puzzled some animal-watchers, who argue that no predator could be *that* clever – so why should there be these amazing refinements which may include such things as small patches of fake leaf-mould, or tiny pieces of 'leaf damage'? The answer is that these deceptions have to fool all possible predators and while some may be able to tune in to one detail others may have a searching-image for something quite different. In the end, the only system that works well, across the board, is the one that involves perfect imitation.

One last camouflage technique that deserves a brief mention is what

In tropical seas a few fish, such as this leafy sea dragon, have evolved bizarre extensions of their bodies, transforming them to look like small floating pieces of seaweed.

The tell-tale clue for many a predator is the circular eye of its prey. Because of this, certain species, such as this banded butterfly fish, possess an eye-stripe, a dark line that passes through the eye and hides its presence.

The most elementary form of camouflage is to match the background perfectly. When the background is a single colour, this is a simple matter, as it is for this yellow crab spider crouching in wait for a prey to arrive on the marigold flower. In the polar regions many species, such as this ptarmigan, have developed a pure white covering.

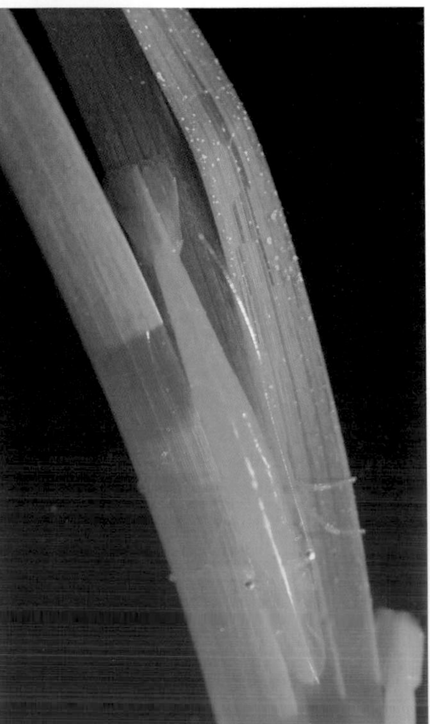

The Mimicry Principle – a tawny frogmouth as a tree-stump and a green shrimp as a plant stem. By copying some detail of their environment many animals become almost impossible for predators to detect.

Half-hiding is a technique adopted by many desert-dwelling species, such as this Peringuey's desert adder. The animal wriggles its body beneath the surface until only the top of the head is exposed. Then, with a watchful eye, it can keep a lookout for potential prey without revealing its presence.

When the pale trees on which the peppered moth rested became dirty from the grime of the Industrial Revolution, only the black variants survived well. These quickly became the norm. This small evolutionary step of 'industrial melanism' was taken in a mere century.

might be called 'half-hiding'. Instead of fleeing to a safe place and hiding away there with complete body concealment, the animal takes up a semi-hidden position from which it can remain watchful. Certain snakes and lizards are highly efficient at wriggling themselves down into the sand or gravel in such a way that only the tops of their heads and their eyes remain exposed to view. With the rest of their bodies submerged, their eyes go unnoticed. Lying completely still they can see without being seen. Among fish this is a favoured technique with many bottom-dwelling species. As they land on the seabed, flatfish flap up the sand in a sudden flurry of rapid movements, and then quickly lie perfectly still. The sand cloud settles on their bodies and obliterates their tell-tale outlines. With a few deft eye-blinks they can clear their field of vision and watch the world about them for clues of special danger.

Some marine species, such as the trumpet fish, hide in a different way. They are lurkers that escape detection by joining the passing throng. As a shoal of some other kind of fish swims past, the predatory trumpet fish moves into their midst and progresses with them as they make their way around the coral gardens. By this simple device, it is able to conceal its highly characteristic body shape and is able to approach more easily its intended prey. For predators make just as much use of camouflage as do their prey. Whether hunting or hunted, the act of concealment is a major survival technique for literally thousands of species.

Warning Signals

The vivid markings and bright colours of these caterpillars (above) act as warning signals telling predators that the potential victims are unpalatable, to be avoided at all costs. The more striking the colours, the easier they are to remember – no predator is likely to forget these particular insects. The markings of this sea slug (below) operate in the same way.

The poisonous skin of these salamanders (opposite) protects them from predation; their bright warning colours help by alerting would-be killers before any damaging bite has been inflicted.

THE MAJORITY OF ANIMALS are not brightly coloured. Their dull coats make them inconspicuous and they survive by avoiding trouble. Those that break this rule and display bright colours and vivid patterns are taking a risk. By making themselves obvious to predators they can easily attract unwanted attention. So for the animal-watcher it becomes a challenge, with every brightly coloured species encountered, to find out what particular advantage there is in giving up the quiet life and facing daily exposure.

There appear to be three main answers. First, the animal may need bright colours for social reasons – to attract a mate, repel a rival, maintain high status, or carry out parental duties. Second, it may need to make itself distinct from other species with which it shares its living space. If that space is a particularly rich environment, such as a coral reef, and food of many kinds is plentiful, many different species will be able to live there together providing they can tell one another apart instantly. Bright colours and complex patterns help to make this possible.

Third, an animal may be brightly coloured as a warning to its enemies that it is not to be trifled with. Any species with a 'secret weapon' – an obnoxious smell, foul-tasting flesh, poison glands, a painful sting or a venomous bite – can advertise the fact by wearing conspicuous warning coloration. Patterns of colours used in this way are called 'aposematic' (the root words *apo sema* meaning, literally, 'away-signal').

We know that these 'go away' signals work because time and again predators have been observed to avoid attacking brightly coloured species. When monkeys were offered hundreds of different kinds of insects, many brightly coloured and many others with dull patterns and colours, only one in five of the bright ones were eaten, while three out of four of the dull ones were eagerly gobbled up. The monkeys were cautious when confronted with conspicuous markings. They may have acquired this caution by observational learning – watching other monkeys trying to eat these insects and seeing the way they reacted to the prey's 'chemical warfare' – or they may have evolved an inborn reaction towards bright patterns, enabling them to avoid such species from birth without any learning process, or they may have learned caution the hard way, by personal experience.

Personal experience, if it happened, would have left them with a deeply embedded memory of an acutely unpleasant incident. If, as naïve young hunters, they attacked a brightly coloured prey, bit it and started to chew it, only to discover that it had a foul taste or a poisonous secretion, they would probably remember it for the rest of their lives. Frequently 'one-trial learning', as it is called, is all that is necessary. Any human being who has eaten a bad oyster, spent a night in agony, and vowed never to touch oysters again, will understand this.

For all its speed, the hunter's learning process is not particularly pleasant for the prey animal providing the education. At best it will be severely prodded and nipped. More likely, it will be mauled and mangled. And in all possibility it will never recover. This is one of the weaknesses of the system. Warning colours work well with educated predators but do involve repeated sacrifices as young hunters learn their trade. Nevertheless, taken on balance, if you possess unpleasant chemical properties, it pays to advertise. Once the learning has taken place, the predators leave you alone and you can go about your business in a quietly relaxed manner, without too much fleeing, hiding, panic, or scuttling for cover. In fact, the pace of your life will be slower and your lifespan may even be extended. Animals with warning colours on average live longer than those without, and there is a simple reason for this. Once they have completed their reproductive life they can live on past it as 'trainers' for novice predators. If they are fatally damaged when attacked, their deaths do not affect the

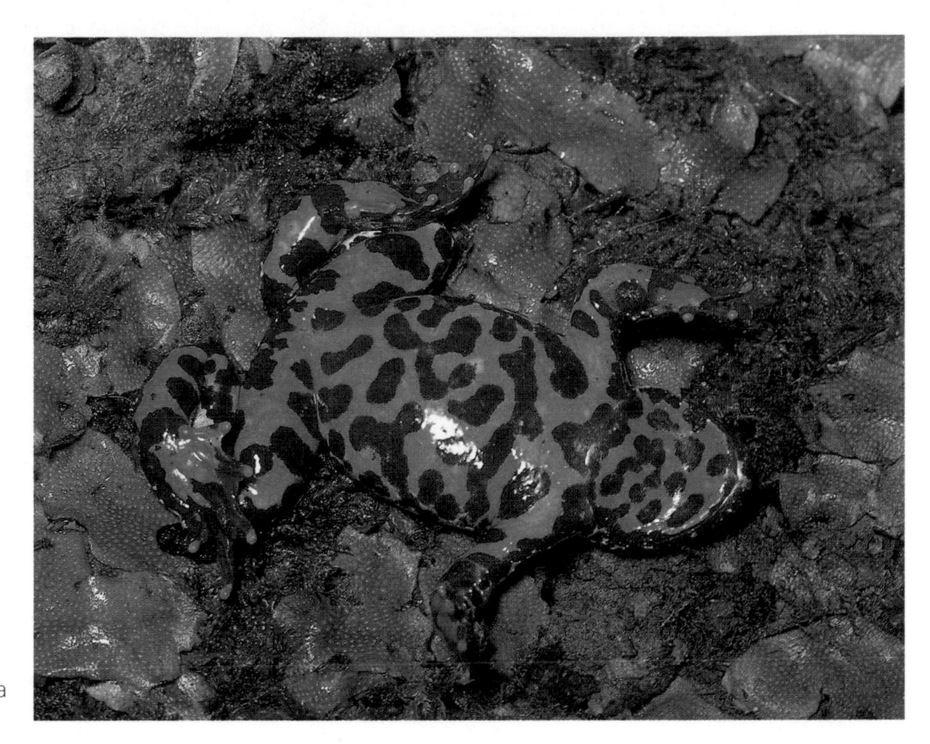

The Oriental fire-bellied toad (right) keeps its bright colours on its underside. They remain hidden until it is threatened, when it displays them by turning itself upside-down. This katydid (opposite), a kind of grasshopper, stands on its head to display its own version of a warning display.

population as a whole because they have already bred and are no longer doing so. They therefore take some of the strain off the younger breeding stock within their species. If dull, cryptic animals live on after their reproductive phase is over they, by contrast, only serve to increase the population density of their kind and thereby increase the chances of a predator finding them, despite their camouflage. Once the killer has stumbled upon one cryptic individual, it quickly develops a 'searching image' and then finds others more easily. So it pays for cryptic species to die soon after their breeding period is over.

One disadvantage of warning colours is that they may attract killers who might have missed the prey animals altogether if they had not been so gaudy. A refinement that overcomes this is a coloration that appears inconspicuous at a distance but extremely bright and vivid near by – an improvement found in certain caterpillars and moths, for example. An alternative is a special posture that the prey adopts when approached by the predator, suddenly showing off its warning colours. The normal posture hides these bright colours; the display reveals them at close quarters. Many amphibians employ this device.

Other weaknesses of the system are that stupid killers do not learn quickly enough, if at all. Predators from the lower ranks of the animal kingdom sometimes seem to be incapable of inhibiting their automatic responses to the brightly coloured individuals they encounter. With each attack, it is as if they are meeting the distasteful animal for the first time. This means that for certain prey being conspicuous is a constant risk and not one worth taking. Only wise predators can be broken in. Also, any killers, wise or stupid, can be driven to eat anything when hungry enough. Starvation overcomes all, and the brightly coloured succumb along with the cryptic. Finally, there are a few killers that have been able to develop a special tolerance for certain poisons, enabling them to eat the noxious prey without ill-effects.

Clearly, though, despite these few shortcomings, the warning display is a useful one. It is found over a wide range of species, from butterflies and wasps to sea snakes and skunks. The yellow and black banding of bees and wasps is easily remembered and avoided. The black and white stripes of

If harmful insects can protect themselves from attack by displaying special markings that warn away predators, then it follows that harmless insects can benefit by mimicking them. Here a stinging wasp (above, left) is imitated by a harmless hoverfly (above, right), a distasteful butterfly (opposite, left, top) has a harmless copy (opposite, left, bottom) – both are from Nigeria – and a venomous coral snake (opposite, right, top) has a harmless mimic (opposite, right, bottom) with almost identical markings. This is called Batesian Mimicry after its discoverer, Henry Bates.

the skunks act as a powerful deterrent, even from a great distance. Warning colours are found among caterpillars and butterflies, ladybirds and ground beetles, frogs and salamanders, and many other animal groups. From close scrutiny, it soon emerges that they all have several features in common. To start with, nearly all the warning patterns consist of pale bands or light patches against black backgrounds. The pale colour is usually white or yellow or orange and shows up vividly against the dark areas. This is not arbitrary – it is designed so that it operates just as well with colour-blind predators as with those that have good colour vision.

Another strange quality of warning coloration is that it varies remarkably little from one species to the next. Even quite distinct groups share the same sorts of patterns. Again, this is no accident and it has a special advantage. For example, if there are several quite unrelated species of unpalatable butterflies living in one particular region and one of them has evolved a warning pattern of black, yellow and red markings, it will pay the others to follow suit because they will then all share the 'training risks' associated with novice predators. Once a young killer has attacked one individual with this wing pattern, it will not only avoid other members of that species but also members of other species that look superficially like it. These 'warning clubs' were discovered by a German field-worker called Fritz Müller. It was as if each species of inedible butterfly was mimicked by the others, and this convergence of colour pattern became known as *Müllerian Mimicry*.

Another form of mimicry was discovered by the English naturalist Henry Bates. Also a field observer of insects, he noticed that many edible butterflies had the same bright markings as inedible ones. He saw that these tasty species were treated with respect by predators as if they too were distasteful to eat. In other words, they were sheep in wolf's clothing. They were the harmless copying the harmful and benefiting from the deception. This type of warning signal is today known as *Batesian Mimicry*. It only works well, of course, if there are sufficient numbers of the genuinely distasteful individuals around to keep the predators well trained. If the mimics swamp out their models, the trick no longer succeeds. Studies with starlings as predators showed that the mimic insects could not rise above the 60 per cent level without rapidly losing protection. But as long as 40 per cent or more of the brightly coloured insects were genuinely distasteful, the mimicry was highly efficient.

Where the prey are merely distasteful or unpalatable, the predators can learn their lesson safely. They may suffer sickness, vomiting or acute pain, but they do not die. They live on, now fully educated, and fill the environment with cautious killers. But what if the prey is itself a killer? What if its poison is lethal? In such cases each novice predator that makes a mistake becomes a corpse and learns nothing. Both the prey and the predators keep on suffering. In such cases it is essential for the predators to evolve *inborn* reactions to the warning signals of the lethal prey, so that they can be avoided without any learning process taking place. One example of this concerns the deadly sea snake and a variety of predatory fish. The brightly marked sea snakes are among the most venomous snakes to be found anywhere in the world. They are confined to the warm waters of the Indo-Pacific region but are absent altogether from the Atlantic Ocean. In a series of tests it was discovered that large predatory fish from the Pacific avoided sea snakes when they were offered as food, but that fish from the Atlantic lacked this caution and attacked them. Ten species of Pacific fish were used and nine from the Atlantic. It is highly unlikely that any of the Pacific fish used in these tests had ever encountered a sea snake previously and in two instances they certainly could not have done so. It appears that in this case, at least, the avoidance of the deadly prey is inborn. Doubtless other such cases also exist.

Chemical Defence

The horrors of chemical warfare are not confined to the laboratories of military scientists. The natural world is full of them. We all know of the dramatic effects of the venoms of snakes and spiders, the savage stings of wasps and jellyfish and the virulent poisons of scorpions and stingrays, but there are many other examples in the animal kingdom and some of them are even more lethal.

The champion poisoners – the Borgias of the forest – are the tiny kokoi frogs that patter boldly about in the tropical undergrowth of South America. These beautifully coloured amphibians, usually less than 2 inches long, carry in their skin-glands a poison so powerful that it is hard to describe its strength. If you collected as little as 1 gram of this toxin it would be enough to kill a hundred thousand average-sized men. If you gathered 1 ounce of it, you could wipe out more than two and a half million people. It is the most potent poison known to us from the entire animal world.

These lethal little amphibians belong to a family that has been given the name of poison-arrow frogs. This is because, long before white men arrived on the scene, the local Indians living deep in the Amazonian rain forests caught and killed these animals to provide themselves with the poison tips for their arrows. Impaled on a stick and held over a fire, the dying frogs exuded a milky secretion. The arrows were held in this liquid and then allowed to dry. One frog was used to coat fifty arrowheads. When the Indians went hunting, whether for animals or for rival Indians, their firepower was deadly. Anything struck by a poisoned arrow was paralysed almost immediately and in a few moments was dead.

Because of their lethal defence system, these small frogs are able to live bold, untroubled lives on the forest floor, hopping about conspicuously with bright colours advertising their presence. Many other animals that harbour secret chemical defences also show themselves fearlessly to the world and some of them also perform special displays that indicate their readiness to engage in a chemical encounter.

Anyone who has encountered a skunk will know that before it fires its stinking spray it issues clear warnings of its intentions. It starts by stamping its feet on the ground and arching its back. If this, combined with its vivid display of black and white fur, does not deter the enemy, it may then start swaying its body back and forth as if cranking itself up for action. At this point it may – especially if it is a spotted skunk – perform a handstand, rearing on to its front legs with its hind legs held high in the air. This amazing posture displays its markings even more conspicuously and in this position it may advance several feet towards the attacker.

In origin, this inverted posture seems to be related to the scent-marking action of certain small carnivores that deposit their odours upwards on high objects. Some mongooses, for example, employ a similar handstand when marking overhanging branches with their personal scent, and South American bush dogs regularly stand on their front feet when leaving their personal odours on tree-trunks. Presumably it is because of the link between standing up in this way and the act of squeezing scent from the anal region that the posture has become associated with the firing of the anal glands by the defensive skunk. The message of the stance is clear enough – come any closer and I will squeeze my glands tight.

However, before actually firing its spray, the skunk must revert to a position in which all four feet are on the ground, the tail is raised and the back fully arched. Then, looking back over its shoulder at the enemy behind it, it aims and shoots. As it does so, it swings its body slightly from side to side, like a machine-gunner raking the enemy ranks. This gives its nauseating, pungent spray a wider range – an arc of about 45 degrees – and greatly increases the defender's chances of covering its target.

Any animal that has been sprayed just once will respond immediately

The most virulent poison known from the animal world comes from the skin of the small, brightly coloured poison-arrow frogs from South American forests (opposite and below).

The skunk is capable of ejecting a spray of foul-smelling liquid from its anal glands, but it always gives fair warning before attacking in this way.

and back away fast from a displaying skunk. And with good reason. The spray is not merely unpleasant, but is also extremely painful if it reaches its target directly. The noxious substance contained in the two anal glands, one on either side of the animal's rectum, is called *butyl mercaptan* and it plays havoc with a predator's face. In addition to the foul stench it produces, it is also damaging to sensitive mucous membranes. If it gets into the predator's nose it can injure the tissues there and if it enters the mouth it causes violent vomiting. If the spray hits the eyes there is almost unbearable pain and temporary blindness. It is pitiful to see the agony of a naïve dog that has nosed up too close to a skunk and been sprayed in the face. Retching, vomiting and desperately trying to rub its face clean, the animal learns a lesson it never forgets.

The stench of the skunk seems to last for ever. The amber-coloured fluid is ejected with great force a distance of between 8 and 10 feet. And the odour can be carried as much as a mile away if the wind is blowing in that direction. Gloves used to handle a defensive skunk still stink of the animal's scent more than a year after contact.

If the skunk has not squirted its glands for some time, then it is primed for about six powerful shots at the enemy, but it is unlikely to use all its ammunition at once because it will take about a week for replenishment and the animal cannot afford to risk being cornered empty-glanded by a second predator. So it uses its chemical defence sparingly. Even so, it is powerful enough to keep almost any enemy at bay.

There are nine species of skunks with a combined range that covers nearly all of North, Central and South America. Their equivalent in Africa is the zorilla, but its chemical impact is not quite so impressive. Several other small carnivores also emit unpleasant stenches when caught by predators and sometimes these odours may deter their attackers. Some foxes, mongooses and weasels are capable of this type of defence and doubtless give the hungry predators the sensation that perhaps, after all, the food object they are pursuing is not so tasty.

In all these cases, the unpleasant chemicals employed are modifications of personal scent secretions, but another major source of chemical defence substances is the saliva. The only mammals with a venomous bite are certain moles and shrews. They employ their poisonous saliva primarily as a weapon against their prey. When a venomous mole or shrew bites its prey, its saliva acts as a paralytic agent. In this way it is able, if it wishes, to make large stores of food. One mole-nest was found with an adjoining store of no fewer than 1,280 worms. The poisonous saliva can be used in a secondary way to deter attackers. In some species the venom is remarkably strong. One American short-tailed shrew stores enough venom to kill two hundred mice.

The only other type of mammal to employ chemical warfare is the duck-billed platypus. On the backs of the ankles of the male platypus there are horny spurs. These are hollow and are connected to poison glands further up the legs. Normally covered by a fold of skin, these spurs can be erected if the animal is caught and, with a sudden jab of the hind legs, the sharp points of spurs are stabbed into the attacker's flesh. The poison injected is enough to kill any dog-sized animal, and even fully-grown humans are at risk, suffering intense pain. The mystery with the platypus venom is why it evolved. The species does not seem to have any appropriate indigenous enemies against which to use it. As it occurs only in the males it has been suggested that it is somehow employed in male-to-male rivalry. Alternatively it could have developed as another example of a prey-killing poison that has been secondarily used against attackers. Certainly when the dingos and their human companions first arrived in Australia the poison-spurs of the platypus found a valuable use as defensive weapons against the new invaders.

The male of the duck-billed platypus has sharp poison-spurs on its hind legs.

Some moles and shrews carry chemical weapons. This short-tailed shrew can inflict an unpleasant bite with the help of spittle from its venomous salivary glands.

The most notorious poisoners of the animal world are certainly the snakes. Their fame in this regard has been their undoing and has caused the needless persecution and destruction of countless millions of harmless reptiles. Because some snakes are genuinely dangerous to man, all snakes are feared, and yet only 15 per cent of the living species are in fact venomous. There are 2,300 harmless species of snakes and only 400 poisonous ones, but few people take the trouble to find out which is which. If a snake is encountered no one takes any chances.

One simple way of spotting a poisonous species is to look at the shape of the head. If it is unusually wide, it is almost certainly a highly venomous form. This is because the poison is stored in greatly enlarged and modified salivary glands. These glands are connected to long, pointed, hollow teeth that act as hypodermic needles, injecting the lethal liquid into the victim's flesh. In the most advanced types these fangs are hinged. When the snake's mouth is shut they lie flat, but when the jaws are opened wide to strike, the two upper teeth that supply the venom are pivoted down through about 90 degrees, so that they are at right angles to the top of the head. This means that when the jaws gape fully open, the sharp fangs are aimed directly forward. At this moment the snake strikes, flinging its neck towards its victim at the astonishing speed of 8 feet per second. The actual distance covered will probably be no more than 2 feet, so it is necessary to respond in less than a quarter of a second in order to avoid being struck.

As the teeth sink into the flesh, the pressure of this action squeezes venom out of the glands and down the hollow tubes of the paired fangs. It spreads quickly once inside the victim and is soon being circulated in the bloodstream. For human victims, the classic he-man response to this injury is the 'cut-and-suck' treatment, in which the unfortunate sufferer undergoes further trauma by having the site of the venom injection sliced open with a sharp knife and the blood sucked from the now gaping wound.

(Opposite) Venomous snakes do not bite their victims – they stab them with open mouths, injecting their venom through hollow teeth in the upper jaw. The fangs of this pygmy rattlesnake are clearly visible but are not yet in the full striking position, which would entail a much wider gaping of the jaws to point the tips of the fangs directly forward.

All this achieves in reality is increased shock and the greater chance of a stress death. It does nothing to reduce the damage because by the time the sucking has taken place the poison is already busily circulating through the victim's body. The best hope is to keep the patient quiet and inject the appropriate anti-venin. Unfortunately it is crucial to know which species of snake is involved because snake poisons are so complex that it has proved impossible to develop a 'multi-venin' that will treat all forms of snakebite. The bite of each species must be treated with the anti-venin specific to that kind of snake. A mistake with the treatment will simply mean a double threat to the life of the patient.

So the typical snakebite victim is stabbed rather than bitten. There is no grabbing, but merely hitting with an open mouth. The reason for this is that the snake must do its best to protect its precious fangs. If it bit its victim and then hung on tight, the ensuing struggle while the injected individual died would probably damage the long, delicate teeth, ripping them from their sockets. So the snake makes its deft, rapid lunge and then immediately withdraws to await results. The poison works so quickly that the victim rapidly succumbs and the snake is then safe to make its move. The primary function of snake venom is, of course, to quieten prey before swallowing them. Its use in self-defence is entirely secondary and only employed as a last resort. All snakes prefer to retreat from predators as quickly as possible, but if cornered will then use their lightning strike as a form of protection. If the victim is not edible the reptile quickly slithers away into hiding, but if it is a suitable prey it simply sits back and waits for the animal's agonies to cease. Then, as it lies dying, the snake moves in and starts to swallow the now helpless creature. In this way the fangs receive the minimum of damage. Even so, they do sometimes become dislodged by the ferocity of the strike. When this happens a new fang quickly grows to replace the lost one. At any moment there are about six fangs in reserve on each side of the mouth, at various stages of development. When an active fang is ripped out, the next most mature one takes its place and quickly develops so that it is ready to fire. The most impressive of all fangs are those of the gaboon viper from Africa. Many of the big vipers possess fangs that are about an inch long, but the gaboon's huge teeth are nearly 2 inches in length, penetrating well into the deeper tissues where there is a rich supply of blood vessels to carry the poison away.

The poison itself is a yellowish, cloudy liquid. It contains neurotoxins (nerve-poisons) and haemotoxins (blood-poisons). In the venom of cobras, mambas and sea-snakes, the nerve-poisons are dominant; in vipers and rattlesnakes the blood-poisons are more active. With a cobra bite there is a creeping paralysis accompanied by nausea and vomiting, leading to convulsions and the cessation of breathing. With a rattlesnake bite there is massive swelling around the bitten part with the flesh turning blue, green, purple or black, with livid blotches and blisters, and this gradually spreads throughout most of the body, affecting the heart and eventually stopping its action. The body of the adult human, however, can often withstand this chemical onslaught and ultimately recover fully. It is large enough to absorb the damage in a way that the smaller, prey species cannot manage. It has been estimated that, of the thousand people bitten by rattlesnakes annually in the United States only about thirty die. This puts the chances of survival at about 33 to 1, and some authorities think this should be even higher – as much as 50 to 1.

It is difficult to obtain accurate figures from the wilder, more snake-ridden parts of the world, such as Africa, Asia and South America, but an attempt at a global survey carried out in the 1950s gave a total world figure of thirty thousand deaths annually from all forms of snakebite.

Perhaps the most remarkable of all serpentine chemical defences is that employed by the spitting cobras of Africa. These snakes have evolved

The spitting cobras of Africa are capable of shooting their venom at an enemy from a distance of 6 feet, causing eye damage and severe pain.

The toad possesses a dangerous poison in its large warts. When the toad is attacked, this chemical weapon oozes out as a thick, creamy liquid. Any mammal biting the slow-moving toad is soon retching and vomiting. Dogs are particularly vulnerable to this poison, but some snakes appear to be immune.

the ability to rear up and squeeze their poison glands so forcibly that the venom inside them is propelled towards their enemy as a jet or spray of droplets. They are capable of aiming accurately enough to splash a man's face from a distance of 6 feet. If the venom strikes his eyes it can temporarily blind him and possibly even permanently damage his sight. The pain is severe and no predator would risk a second close encounter with these snakes.

Unlike snakes, lizards have not specialised in chemical warfare. Of the three thousand living species only two possess venom. They are the gila monster and the beaded lizard, from Mexico and the southern states of the USA. Their poison has the same origin as that of the snakes – modified saliva – but the animals are much more primitive in their wounding mechanism. There is no hypodermic injection. A lizard simply holds tight with its teeth. The venom trickles into the enemy's damaged flesh as the lizard slobbers and struggles to keep its grip. Like a bulldog it refuses to let go, even when savagely attacked by the desperate victim. But it would be wrong to give the impression that these are aggressive animals. They avoid trouble at all times and only if they are cornered or pestered will they bite at their attackers. Their neurotoxin produces effects similar to those seen in snakebite victims, although the number of deaths from lizard-bites is extremely small. Indeed, only eight cases have been fully authenticated.

Among the amphibians there are many with poisonous skins, but none that has a poisonous bite. If you are to suffer from amphibian poisoning, it is you who must do the biting. The deadly little poison-arrow frogs have already been mentioned, but there are many other frogs and toads whose

There are only two kinds of poisonous lizards, the gila monster, seen here, and the similar but larger beaded lizard. Both bite their attackers and hang on like bulldogs, letting the venom trickle into the wound as they do so.

The most common circumstance in which human swimmers are poisoned by fish is when a well camouflaged stingray is trodden on. It reacts automatically by lashing upwards with its long thin tail and stabbing the victim's leg or foot with the sharp spine near its tip. This spine has a serrated edge and can inflict painful wounds.

kin glands and mucous secretions can cause vomiting and nausea in redators that grab hold of them in their jaws. Almost all these poisons re derived from the various substances that have evolved as lubricants or the soft, vulnerable skins of these small animals. Where the mucous as become distasteful, the animals bearing it have prospered, and n this way increasingly distasteful and eventually highly poisonous ecretions have been favoured by natural selection.

The world of fish is full of needle-sharp spines that can inject agonising oisons into unwary flesh. The greatest threat to humans comes from the tingrays, large flatfish that lie inconspicuously on the seabed in shallow vaters. If a human foot accidentally treads on one of these shy, naggressive animals, it responds instantly with an upward flick of its ong, thin tail. Towards the tip of this tail there is a sharp spine with a errated edge and this is slashed painfully against the offending foot or leg. ll along the underside of this spine there are little grooves lined with a pongy venomous tissue. The tissue contains soluble proteins that cause ntense pain and swelling, but which are rarely lethal to man. Out of a housand cases reported in United States waters over a five-year period, nly two resulted in death.

More deadly and much more painful is the chemical weaponry of the eavily camouflaged and lethargic stonefish. For a human swimmer, to tep on one of these almost invisible bottom-dwellers is to experience erhaps the worst possible pain. Even the most fiendish human torturer as fallen far short of the agony provided by this small, shallow-water fish. s only response to being disturbed is to erect its dorsal spines. It makes no ttempt to move away, even if the water near by is being violently isturbed. The venom is held in small sacs near the tips of the spines, so at when the human foot descends on the fish the poison is squeezed into e wound by the downward pressure. The fish does nothing – the victim npales himself. And the ensuing pain is so intense that victims seem, for a hile, to go out of their minds. They rush from the water, fling themselves own on the beach and proceed to roll around in agony, screaming and othing at the mouth. Helpers approaching have sometimes been bitten or tacked wildly in the delirium that follows and which may last as long as

At first sight the cone shell does not appear to be particularly dangerous but it carries a secret weapon that can cause an unusually painful death. When attacking, it protrudes its long snout, and as it pushes it against the flesh of its enemy it fires a small, venom-carrying harpoon – a hollow dart formed from a tooth – that can kill a human being within a few hours.

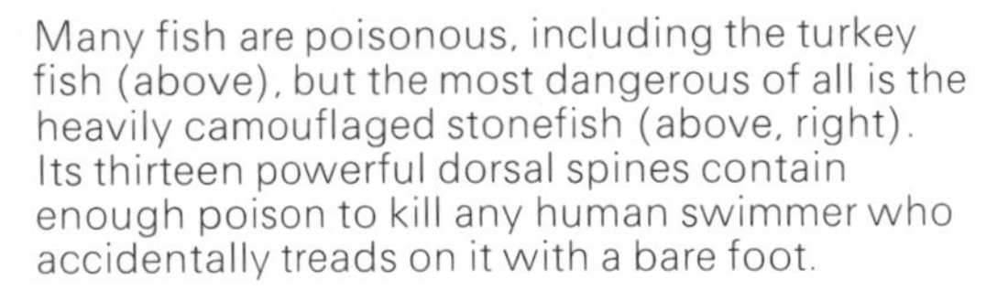

Many fish are poisonous, including the turkey fish (above), but the most dangerous of all is the heavily camouflaged stonefish (above, right). Its thirteen powerful dorsal spines contain enough poison to kill any human swimmer who accidentally treads on it with a bare foot.

twelve tormented hours. Morphine has no effect. Raving and thrashing about, the victim becomes gradually weaker and, if lucky, may eventually become unconscious. The injured limb swells up until it looks like the leg of an elephant. The toes or fingers nearest the wound turn black and later drop off. Death is not uncommon, but an anti-venin has now been produced and is effective if it can be used in time.

The stonefish is the most deadly of all fishes, but many others possess poisonous spines developed from fin-rays, and any prickly-looking fish should be treated with great respect. Other examples include the scorpion fish, lion fish, rabbitfish, toad fish, waspfish, frogfish and many of the catfish. Some, like the stonefish, are inconspicuous, while others flaunt their chemical power by sailing around in bright colours that dare you to touch them. The beautiful lion fish belongs to this gaudy category and is therefore much easier to avoid. All these spikes evolved originally as a way of preventing predation by larger fish. Today, with powerful venom added, they have become a formidable defence system against anything that enters the domain of the spiny fish.

Also infesting this domain are the dramatically spiny sea urchins. Some species possess poison bags near the tips of the long, brittle spines. When trodden on, these spines puncture flesh and usually break off inside the resulting wounds, ensuring the prolonged insertion of the painful poisons.

Surprisingly, certain sea snails are also quite deadly. The seemingly harmless cone shells, attractively patterned and very slow-moving, can kill with their sting and human medicine has to date found no antidote. A harpoon-shaped tooth is thrust into the flesh of the victim, carrying the venom with it. Numbness, paralysis and eventually heart failure result. But the only way one of these snails will attack is if its soft parts are manhandled. Holding it by the hard shell is safe enough, although a steady grip is recommended.

More difficult to avoid are the poisonous jellyfish. As they float through the water their long, almost invisible tentacles hang down and spread out far from the visible central part of the body. In some species these trailing threads, full of stinging-cells called nematocysts, can reach as far as 50 feet. Swimming through a jellyfish-infested sea is like advancing through a minefield. Each stinging-cell contains a coiled-up tube. If a victim brushes up against a tentacle, literally thousands of these tiny cells automatically fire, shooting out the sharp-tipped tubes like minute harpoons. These penetrate the skin and paralyse the intruder. If the victim is a small fish paralysis and death are rapid, and it can then be drawn in and eaten. But

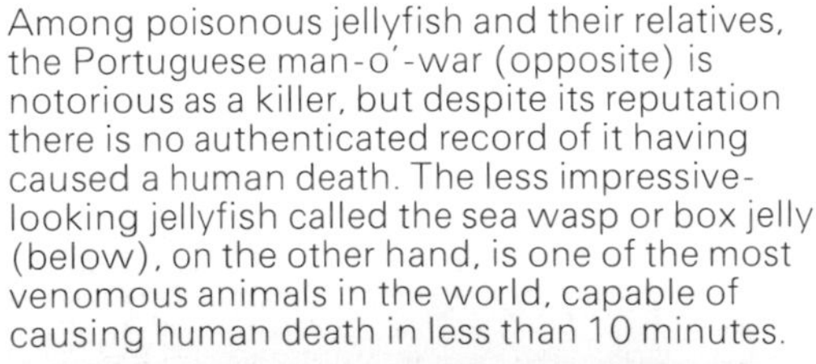

Among poisonous jellyfish and their relatives, the Portuguese man-o'-war (opposite) is notorious as a killer, but despite its reputation there is no authenticated record of it having caused a human death. The less impressive-looking jellyfish called the sea wasp or box jelly (below), on the other hand, is one of the most venomous animals in the world, capable of causing human death in less than 10 minutes.

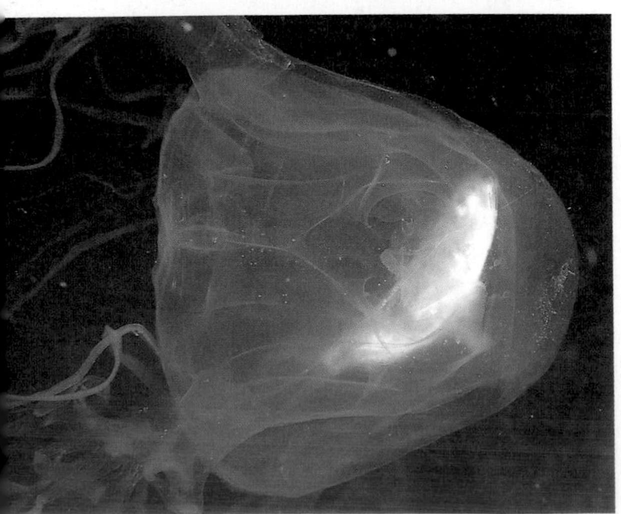

the body struck by the stinging-cells is large – such as a human swimmer's – the poison is usually insufficient to kill. Instead there is searing pain, blistering of the skin and a rapidly discovered aversion to swimming in the sea. The most virulent of all jellyfish is not, as is usually claimed, the large Portuguese man-o'-war, but the sea wasp or box jelly commonly found off the shores of Australia. The neurotoxins contained in its thousands of little venomous darts are strong enough to cause both extreme pain and sometimes even death. Since it stings anything it bumps into, regardless of what that may be, the only solution is to stay clear of the water when the swarms of jellyfish appear.

Staying clear of water is no protection against the hundreds of poisonous insects, spiders and scorpions that abound on dry land. Some sting, some bite, some throw poisonous hairs at you and others squirt you with blinding sprays. Most are unpleasant but not lethal. A few are deadly.

The most dangerous of all the spiders is the black widow, a small species that is common throughout America and which responds to vibrations of its web by rushing out and biting what it considers to be its next meal. Its bite is rarely lethal, but is so painful that some victims have survived when they 'hoped not to'. A meticulous survey of black widow attacks on humans in the USA from 1726 to 1943 revealed 1,291 recorded bites – in other words, an average of no more than six bites a year. Only 55 caused the death of the victim – or one every four years. Despite its bad reputation, it is clearly not a species to cause concern.

Even less of a threat is the huge tarantula or bird-eating spider. This has the worst reputation of any 'creepy-crawly' in the public imagination, and yet its bite, which is extremely rare, is no worse than a bee sting. Its usual reaction when abused by an attacker is to tear out its own hair rather than attempt to bite. It flings its fine hairs in the face of the assailant, in a dense cloud. When this cloud settles on the skin it causes a red rash, but it is no more than an irritation and certainly nothing to justify the terrible image of this animal that has been so relentlessly fostered by cheap novels and films. As with so many poisonous animals, the tarantula prefers to keep its venom for feeding time and only employs it against enemies as a last resort.

It is, in fact, impossible to find any poisonous animal that actively goes in search of victims unless it is feeding. Despite popular mythology concerning 'evil' and vicious species that track down their victims ruthlessly and mercilessly, the reality is that no venom, no poison, no bite, no sting, is ever administered except under extreme provocation or simply by accident, unless the victim is either a potential prey or a rival belonging to the same species. As in human warfare, chemical defences are essentially deterrents rather than everyday weapons.

The black widow spider is the most dangerous of all spider species and its venomous bite has been known to cause human deaths on rare occasions. But it has been calculated that there is only a 4 per cent chance of dying if bitten by one of them.

Deflection Displays

DEFLECTION DISPLAYS are deceptive. They are examples of natural deceit. Just as the magician fools his audience with his sleight of hand, so deflection displays confuse the pouncing predator by sending him false signals. Their essential feature is that they misdirect the enemy's attack, so that it fails to damage any vital organs.

Most predators prefer to attack their prey at the head end. When zooming in for the kill, or performing a lightning strike, they ignore the tail end of the body and aim straight for the eyes. This targets them towards the most important part of the prey, where they can do most damage with their first contact. Even if the prey is not killed outright, there is a greater chance that it will be incapacitated. Many a prey species can afford to dispense with its tail but can never afford to lose its head.

Also, most prey can only flee fast in a forward direction, so attacking the head is the best way of hampering their escape. And many predators like to grasp the prey by the head to facilitate swallowing. In the case of fish, prickly fins and sharp gill-covers all point backwards. Gulping such a prey animal down head-first is a smooth operation. Swallowing it backwards can be painful or even fatal.

For these reasons, some prey species possess display markings that confuse their heads and tails. They make their heads less head-like and wear false eyes on their tails. The predator no longer knows where to strike.

There are several degrees of deception. The simplest involves displaying a large eye-spot marking somewhere at the rear end of the body. This looks as eye-like to the predator as the real eye, and gives the prey a fifty-fifty chance of losing its head. Usually the false eye is a little bigger than the real one, tipping the odds slightly in favour of the prey. A predator dashing in for the kill has no time to assess the minute details of the eyes to detect whether they are real or false. It simply sees 'eye' and reads 'head'.

A more advanced technique involves camouflaging the real eye while at the same time making the false eye truly spectacular, with a huge black 'pupil' surrounded by a bright white eye-ring. This tips the balance much more in favour of the prey and sends most attackers lurching towards the wrong end of the body. Even if part of the tail is lost, the victim survives and swims away to safety.

An even greater refinement is possible. Two species of butterfly fish have been observed to swim slowly backwards at the first hint of trouble, making their false eyes seem even more real. Then, when the predator moves in for the kill, at the very last moment the butterfly fish switch direction and dash rapidly forward, leaving the frustrated would-be killers snapping at empty water.

Deflection eye-spots are found in a wide variety of quite unrelated species. They seem to have evolved not once but many times. As a result they show subtle differences, some being positioned on the upper fin, some on the rear of the main trunk and some on the tail-stem. The elimination of the real eye is also achieved in various ways. Sometimes there is a single vertical bar running right through it to obscure it. Sometimes the whole body of the fish is banded with vertical marks, one of which conveniently blots out the real eye. In other cases, the eye-concealing lines are horizontal stripes, or a single diagonal line may run up through the eye and end on top of the head. Occasionally, the real eye is hidden in a mass of spots, sometimes white, and sometimes black. In one unique fish, the imperial angelfish, the juveniles are covered in white lines that curve round in a special pattern that centers all one's attention on the tail end. There is no actual eye-spot, but the effect is just as compelling, with the curved lines shrinking to smaller and smaller circles as they approach the tail. The head, by contrast, is obliterated by formless, meandering lines.

A large eye-spot at the tail of a fish deflects the interest of a predator away from the more vulnerable head. Having an 'eye' at each end of its body, this Oscar cichlid gives its predators only a fifty–fifty chance of striking its head.

On this butterfly fish the eye-spot at the rear of the body becomes the centre of attention, the real eye being obscured by a vertical stripe.

This sharp-nosed puffer fish (left) has a real eye, on the right, that is not inconspicuous. Despite this, however, one's gaze is drawn to the huge black false eye at the other end of the body.

Some fish go a step further and display not one but a pair of false eyes. This is even more confusing because the predator receives the impression that it is approaching a fish that is swimming straight towards it. These false eyes stare implacably, challenging the predator to bite, while the small, real eye of the prey is sufficiently concealed not to give the game away.

Fish are not alone in staging this deception. Reptiles and insects employ similar tactics, but in their case eye-spots are not necessary to deflect the interest of their predators. All they need is a patch of bright colour on their tails. Among the lizards, this can be seen in young skinks of several species. Their heads are inconspicuous and camouflaged with black and white stripes that conceal the eyes. These stripes run right down the body, but peter out near the base of the tail. The rest of the tail is bright, vivid blue – so intense that it is hard to see anything else in its vicinity, including the head and body of the small reptile. When approached by a stalking killer, the lizard's tail is the focus of attention. If the predator pounces and manages to catch hold of the tail, it quickly breaks away from the reptile's body and, even in its severed condition, continues to writhe and twist, either on the ground or in the mouth of the triumphant killer. While the lizard scuttles away to safety, the predator is left to discover, little by little, that its apparently succulent meal is in reality no more than a scaly scrag-end.

Most skinks lose this blue colouring as they become adult. Their speed of escape reaction is by then greater and they have become much more experienced. On balance they are better off being fully camouflaged at this stage, because although the blue tail may save the life of many a young lizard it also has the disadvantage of being so conspicuous that it can, to some extent, attract unwanted attention in the first place. In some species that are heavily preyed upon, such as the Polynesian blue-tailed skink, the

The comet (opposite, top) is a fish that attempts to lose its real eye in a sea of white dots. These dots do not, however, conceal the huge eye-spot at the rear end of the body.

The vivid markings of the emperor angelfish (opposite, bottom) create a strong circular motif just in front of the tail. There is no such focus of interest at the head end. Again, a predator's attention is deflected away from the vital head organs.

Some lizards and snakes possess a dull-coloured body combined with a brightly coloured tail. When approached by a predator, the tail attracts all the attention and is more likely to be attacked – the vulnerable head is thus spared damage. In the western skink (below) the tail is bright blue; in the ring-necked snake (right) it is orange.

This owl butterfly (opposite) uses a different deflection device, having a pair of small eye-spots near the trailing edge of its wings. If a bird pecks at these it will probably only make a small tear in the edge of the wing and the butterfly may then escape.

Some butterflies and moths have evolved false heads at the rear ends of their bodies, deflecting the attacks of birds away from their vital organs. This hairstreak butterfly has achieved the effect by evolving not only a coloured false head but also a radiating black and white pattern on its wings that directs the observer's eye away from the real head.

tail colour is unusually retained right through into adulthood. The balance there continues to favour a small tail-end sacrifice even with experienced individuals.

Some snakes go a little further. When approached by a possible killer they hide their real heads and raise up the tips of their tails. These they waggle boldly in the air. In the Malayan pipe snake the tip of the tail is bright red and the sight of this little patch of colour moving about so invitingly is too much for most predators to resist. They attack the tail and ignore the vital head end of the snake. Unlike the blue-tailed lizards, these snakes only have the bright colour on the underside of the tail-tip. This is an improved model, because when the snake is crawling along with its tail on the ground the red colour is concealed. It only becomes conspicuous after an attack has started, and never provokes that attack.

Insects adopt two different devices. Some butterflies have a rear end that looks more like a head than their real heads. The trailing wing-tips have conspicuous head markings and even develop long, fake antennae, while the actual head is inconspicuous. Even more convincing heads are found on some lantern flies, where the real head is well hidden and the false head has a huge eye-spot and bold 'antennae'. These insects complete their deception by jumping backwards when attacked.

The other method, used by certain butterflies, is similar to that found in the deflection-displaying fish. Paired eye-spots on the rear wing-tips give the impression of faces watching. Specimens of these butterflies are often found with bite-marks in the region of these false eyes, showing that bird predators really are attracted to this part of the body, allowing the insect to fly off to safety with no more damage than a few frayed edges.

This type of butterfly eye-spot should not be confused with the much larger eyes that occur on some species – eye-spots that are suddenly flashed at the enemy as it draws close. These do not function as deflection devices but as startle displays. Rather than misdirecting attacks, they repel them altogether, as we shall see in the next chapter.

Startle Displays

The spit and hiss of a cornered cat may startle an attacker because of its suddenness and also because it is reminiscent of the threat display of a venomous snake.

The sudden, unexpected appearance of the bright blue tongue of this skink may startle an approaching predator sufficiently to make it think twice about attacking.

If an animal is approached by a killer it may try to defend itself by giving its attacker a fright. If it can startle the predator in some way, there is a faint chance that the enemy may panic and flee. At the very least, the assailant may be made to pause for a moment before attacking, and this may be just enough to allow the prey to escape. Buying time, whether a lot or a little, during an attack is always worth doing. Many animals attempt this and there are two popular devices: sudden sounds and suddenly displayed fright-patterns.

It is the suddenness that is vital. Nothing gradual can startle anything. As directors of horror movies know so well, there is no better way to scare their audience than to have some unidentified object flash suddenly on to the screen, very close to the hero's face. As the strange shape darts in, the whole audience jumps and gasps. It is a deep-seated defensive mechanism that we share with most higher forms of life. Firing a gun behind an unsuspecting head has the same effect. The body tenses, the neck hunches and the eyes close protectively. It is during such a split second that a prey animal may be able to make its escape, if it is fast enough. If it is a slow-moving species, then it must rely instead on provoking a panic response in the attacker, a response that sees it pull back in horror and retreat to a safe place.

Some mammals, especially those that hide in dark dens or crevices, respond to the approach of a predator with an explosive spit and hiss. The suddenness of the spit makes the attacker jump smartly backwards and the hiss that follows alarms it even more; it retreats carefully, watching the dark corner from where the noise is originating. It is not hard to guess why this sudden display is so effective. Venomous snakes spit and hiss when they are cornered, and no predator likes to risk the possibility that a lurking shape may possess poisonous fangs.

This snake-mimicry is employed by many species, from the familiar domestic cat to less familiar finches. A cat cornered by, say, a dog, will produce an explosive spit that always makes the approaching animal jump back in alarm. As it hisses in its fake-snake manner, following the spit, it completes the illusion by thrashing its tail about in a serpentine way. In the semi-darkness it is a brave predator that ignores this bluff display.

Some small cavity-nesting birds, rather surprisingly, employ similar tactics. A little African finch, the cut-throat finch, when it is disturbed while sitting inside its ball-shape nest, gives a bizarre 'snake-dance' reaction. Anyone who has peered too closely into a cut-throat finch's nest when the bird is sitting on eggs will have experienced the disturbing feeling that, very suddenly, their eye is far too close to a snake for comfort. Instead of fleeing, like most small birds approached this closely, the cut-throat sits firm and hisses like a snake. It then adds a remarkable visual display, writhing its body in a sinuous, undulating movement, exactly like a crawling snake. Even human beings find this distressing to encounter at close quarters without any warning, and find themselves drawing back. Our fear of snakes is as deep-seated as our fear of sudden movements. And the same is true of many other predators. In fact, it is difficult to understand why more nesting birds do not use this defence.

Visual displays that involve the sudden appearance of bright colour patches or conspicuous patterns, especially eye-spots, also intimidate approaching killers. Because poisonous animals are often brightly coloured, as a warning to would-be predators, there is a startle value in a prey species suddenly flashing a bright patch of colour, even if it is not poisonous. This is not simple mimicry, which would only entail being the same bright colour as a distasteful species. It is a combination of mimicry and sudden exposure. The lizard known as the blue-tongued skink is normally inconspicuous. Only when threatened does it flash its huge broad tongue at its assailant. The bright exposure of a vivid blue patch

When the camouflage of this praying mantis fails and a killer is closing in, the insect adopts the last-ditch defence strategy of flashing its wings open and rearing up with its front legs spread. This sudden transformation may help to scare away a timid attacker.

alarms the attacker and may save the lizard's life.

Several kinds of insects adopt a similar strategy. Well camouflaged stick insects, whose concealment has for once failed to work and who are about to be snapped at by a hungry predator, will suddenly flick open their brightly coloured wings in a dramatically unexpected display that may shock the enemy into retreat. Some moths add an extra deterrent. One West African species not only flicks open its wings to reveal a bright red abdomen but also bubbles out a stinking yellow froth from its body.

Perhaps the most effective shock tactic is to create the impression that you are a lethal killer yourself, and not to be trifled with. Many a harmless animal manages to transform itself, with a sudden display, into what appears to be a vicious killer – a scorpion, a snake, or a bird of prey. The camouflage devices employed are truly remarkable. Some caterpillars possess the markings of small venomous snakes and, when approached too closely, will rear up on the branch and wave their fake snake-heads at the attacker. One kind of stick insect arches its abdomen and makes savage swipes through the air as if it is an attacking scorpion. Only an exceptionally hungry or courageous predator will risk attacking such a creature, and most prefer to play safe and leave well alone.

By far the most common form of startle display involves the sudden exposure of a large pair of eyes. It is not always possible to say precisely which kind of eyes these are pretending to be, but the chances are that in most cases they are mimicking the forward-facing eyes of birds of prey. If a killer thinks, even if only for a split second, that it is face to face with an owl or an eagle, it is very likely to back away and this may give the harmless bluffer enough time to escape.

But these eye-spot displays are not confined to harmless species. One of the most deadly animals in the world, the cobra, employs just such a device. As it rears up in self defence, it spreads its wide hood to reveal a massive pair of eye-spots. To at least some of its attackers this pattern must be intimidating and its value is that it may allow the cobra to repel its enemy without resorting to the use of its precious venom. In popular

Predators are usually nervous of attacking venomous species and this stick insect attempts to startle its assailants by adopting a scorpion posture when approached closely. By curving up its abdomen it gives the false impression that it is about to strike with its tail.

Some caterpillars manage to startle their attackers by raising themselves up like small snakes. Aiding them in their deception, their rear ends look remarkably like snake-heads, complete with impressive serpentine eyes. In the case of the citrus swallowtail caterpillar (bottom) there are false eyes near the front end of the body and, just behind the animal's real head, an organ that looks like the forked tongue of a snake. This is bright red and is extruded and flickered at the enemy when the caterpillar is disturbed. Enhancing its startling effect this 'forked tongue' also gives off an unpleasant, pungent odour.

mythology poisonous snakes are always ready and waiting for the chance to strike out and kill their attackers. In reality they will do everything in their power to avoid this, because the primary function of their venom is to incapacitate their prey and they cannot afford to waste it unless they are driven to extremes. So even a lethal cobra prefers to bluff an attacker if given the opportunity.

Less well known than the cobra eye-spot display is a similar action by certain kinds of toads. Toads are poisonous, too, but like the cobra they prefer not to use their poison except as a last resort, although for a different reason – their poison glands only secrete their lethal dose when the toad is bitten hard by an attacker, and for the toad this is leaving matters far too late, with serious injury a strong possibility. The alternative is to scare off the enemy and this is done by inflating the body and raising the rear end of it towards the approaching assailant. There, on its otherwise camouflaged skin, appears a pair of huge staring eye-spots that make its rear view look remarkably like the face of a predatory bird.

Many kinds of butterflies and moths perform similar startling transformations. The peacock butterfly normally rests with its wings pressed tightly together. Their visible undersides are camouflaged, but if the camouflage fails to work and the butterfly is attacked it then immediately

A pair of false eyes is also found on various higher forms of life, including reptiles and birds. Startle eye-spots appear on the hood of the rearing cobra (right), and even on the back of the head of the pearl-spotted owlet (above).

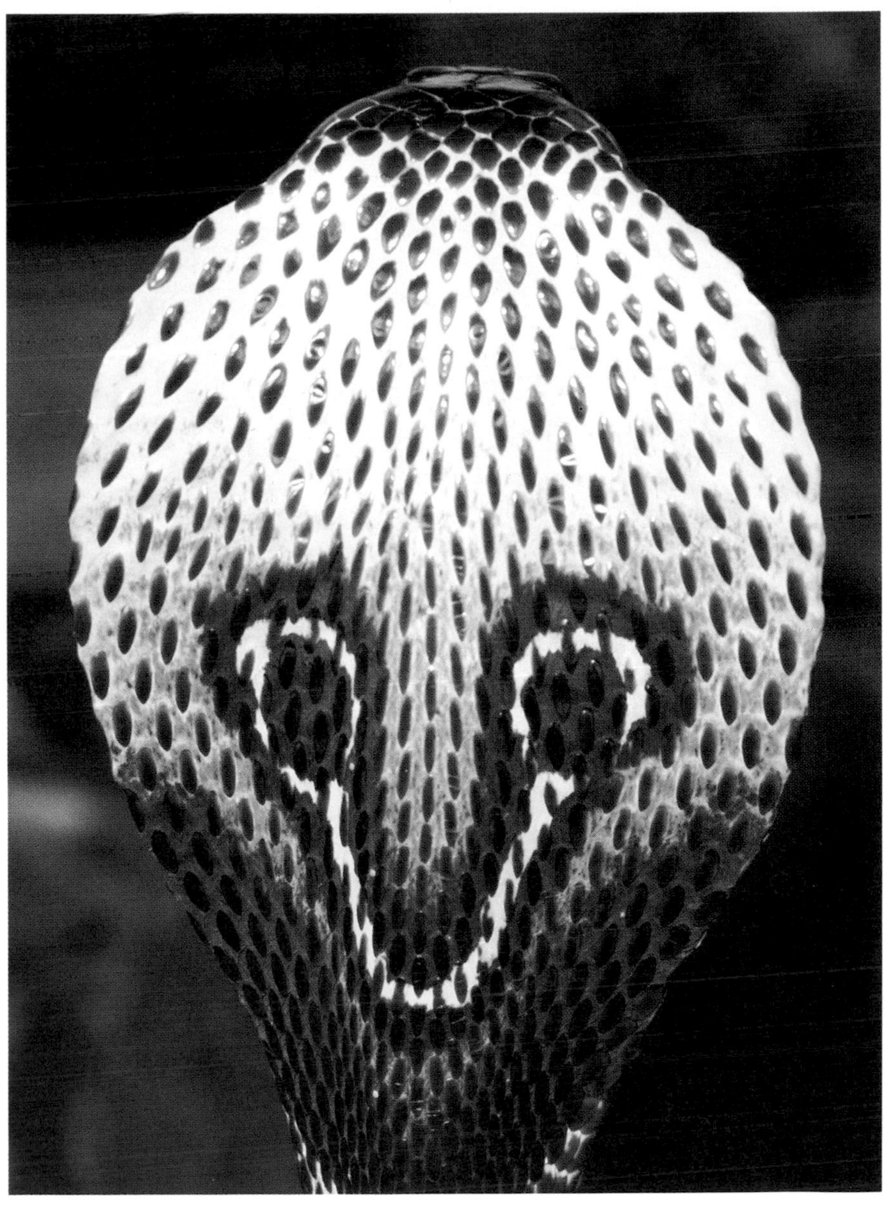

opens its wings flat, revealing two conspicuous pairs of eye-spots. Many moths, like the eyed hawkmoth, have a dramatic pair of eye-spots on their hind wings. They rest with their wings flat, but with the front wings covering their rear ones. Their front wings give them the protection of camouflage but they can be pulled forward rapidly when the moths are attacked, to flash the startling false eyes. An African praying mantis has even more spectacular eye-spots with an almost hypnotic spiral design that it employs in a similar way. In its case, however, the eye-spots are on the front wings and are exposed by suddenly spreading the folded wings and directing them at the attacker.

Most of the enemies of these insects are small birds out searching for a tasty meal. We know that the eye-spots do startle them because of a series of careful field-tests. A special box was used, on to the lid of which could be flashed a variety of patterns whenever a bird approached it. To attract the birds a mealworm was placed in the centre of the box and then, as the bird alighted to attack the mealworm, the box was illuminated to reveal a pair of eye-spots or some other pattern. Using crosses, dots and circles, it was possible to find out which patterns were most effective in frightening away the birds. It established beyond doubt that it was the eye-spots that worked best, and that the more realistic they were the more they startled the birds. In particular, it was found that if the fake eyes were shaded to look three-dimensional they made their biggest impact. This explains why so many of the eye-spots on the bodies of prey animals have evolved such a subtle and complex degree of mimicry. Simple spots will not do, but natural selection has met the challenge and has produced some truly amazing replicas complete even to glinting highlights.

Insects with eye-spot startle displays include this remarkable bush cricket (above). At rest the cricket looks like a dead leaf, but it transforms itself at the last moment.

Many butterflies and moths, such as this io moth, possess a pair of startling eye-spots on their wings. These are usually concealed and then flashed at the last moment, giving the prospective predator the impression that it is approaching a much larger animal and one that is staring straight at it in a defiant manner. Field experiments have proved that this startle display does indeed deter attacks from small birds.

Death-feigning

THE TRICK OF 'PLAYING possum' is not confined to opossums. Acting as though dead has also been observed in a variety of insects, frogs, snakes, birds, squirrels and foxes. In each case the animal, when attacked, suddenly collapses and remains immobile. Limp and apparently lifeless, it permits itself to be mauled and prodded without twitching a muscle. Then, if left alone, it quickly revives and makes a dash for freedom. It is a risky form of self-defence, putting the animal completely at the mercy of its attacker's jaws, but it sometimes works and the bluffer escapes.

Among the insects, death-feigning (or thanatosis, to give it its technical name) occurs in some beetles, bugs, grasshoppers, stick insects and mantids. Certain spiders also use it as a last resort. Giant water bugs may go rigid when taken from the water and may hold this 'frozen' posture for as long as fifteen minutes. Some species of small beetles behave in a similar fashion, pulling all their legs in close to their bodies and lying absolutely still. The larger, longicorn beetles use a modified technique. When approached they may stay absolutely motionless and allow themselves to be picked up. They appear to be in a catatonic state, but this is deceptive because they may then suddenly bite the hand that holds them. This behavioural pattern gives them the chance to avoid detection initially by staying very still. If caught, they can avoid arousing the predator's urge to kill through not wriggling, and then panic the predator into dropping them by making a sudden unexpected attack. Once on the ground they may be able to get to safety before the killer strikes again.

Frogs and toads occasionally employ this device, but it does not seem to be very widespread among these amphibians. Some respond to an attack by drawing their legs tightly against their bodies, closing their eyes and staying completely still. If a frog is climbing when it is attacked, it may drop to the ground and lie there motionless for several minutes before risking an escape. Large bullfrogs have a special reaction. If cornered, they may thrust out their limbs in a fully stretched posture, flatten their bodies and then hold this rigid, spread-eagled pose until danger has passed. They will retain this tonic immobility even if they are placed on their backs, giving certain human tricksters the chance to claim that they are capable of 'hypnotising frogs'.

Chameleons have been known to death-feign, lying inert on their sides, with their feet held out stiffly and their eyes half-closed.

Of all the animals that sham dead it is the snakes that are the star performers, some of them going to amazing lengths to convince their assailants that they are truly deceased. The most extraordinary example of death-feigning in the entire animal world is undoubtedly that shown by the West Indian wood snakes. These are small boas which, if molested, first coil themselves up tightly, then release a foul-smelling fluid from special glands, giving the impression that they are rotting and that advanced putrefaction has already set in. Following this they release blood into their eyes which turns them a dull red colour. But the climax of their act is the breaking open of small capillaries inside the mouth, so that blood is seen trickling from their apparently lifeless jaws. This auto-haemorrhage, as it is called, is an astonishing refinement of the death-feigning strategy, implying that predators pay a great deal of attention to minute details of their prey's condition.

Most snakes do not go this far. Small tree snakes that sham dead do so simply by dropping to the ground when disturbed and lying there. The common European grass snake and the American hog-nosed snake both employ a death-feigning act which, although elaborate, suffers from what might be aptly described as a fatal weakness. They begin well enough, rolling on to their backs, gaping the mouth awkwardly wide and allowing the long tongue to loll lifelessly from their sagging jaws. The impressive realism of this performance is totally ruined, however, if you bend down and turn the snake up the right way again. It responds to this helpful

Some snakes play dead by inverting themselves and lying limply on the ground, but if they are turned over they spoil the effect by quickly inverting themselves again. The death-feigning snake often lies with its jaws gaping open and its tongue hanging out. Then, when the danger has passed, it starts to turn itself slowly back to its normal posture, beginning at the head end. This device is found in a variety of species including the grass snake (right) and the hognose snake (below).

adjustment by promptly inverting itself into the mock-dead posture. Even if the snake is turned over again and again, it always flips itself back into the inverted posture, an over-eager dying act that completely gives the game away. Only when it is totally convinced that the danger is past will the snake finally and with great caution start to move away. It does so by first pulling in its tongue and closing its jaws. Then it rotates just the head end of its body into the normal position. With the rest of its length still inverted, it looks around and tests the air. Then, slowly, very slowly, it starts to slither away. If, as it retreats, it is approached closely once more, it hurriedly flings itself upside-down again and resumes its forlorn death-pose.

Birds rarely employ death-feigning but sometimes a small bird held in the human hand has been seen to collapse motionless and lie very still. If it is closely examined at this point it appears to have died from shock, but then a few moments later it may revive miraculously and fly off. It is a trick that might just save the life of the bird if a cat has caught it and then placed it on the ground for a moment before starting to pluck it. Cats do often pause after catching their prey and momentarily release what they imagine is a corpse before settling down to the business of eating it. So it is possible that avian death-feigning is tuned in to this one crucial moment of possible escape. If the bird keeps struggling, it will be clawed and bitten repeatedly by its captor, but if it lies still the cat may then make the mistake of putting it down on the ground and releasing it from its jaws.

Among the mammals, some South American foxes have been seen to play dead when caught, as have certain squirrels. A little African ground squirrel gave a most convincing performance when examined by a zoologist. The young squirrel had been caught and placed in a box. When the box arrived, the zoologist took the animal out to examine it, only to see it keel over and lie motionless on her hand. No respiration was visible, the eyes were closed, the mouth gaped and the paws were rigid. The zoologist was upset because she felt that the small animal had suffered from the journey in the box and she was responsible for its death. She tried to feed it with a little warm milk, but there was no swallowing response. Putting the 'corpse' down, she gave up and was gazing sadly at the small body when, about two minutes after its 'death', it suddenly took a few very deep breaths and started to move about. It became such a tame pet afterwards that she could never persuade it to play dead again, even if she grabbed it as though she were a predator. It was now too much at home with her for the response to be triggered.

The animal most famous for shamming dead is, of course, the American opossum. It is the species in which this curious behaviour has been studied most thoroughly, with some surprising results. When attacked, the opossum's first reaction is to fight back. It possesses a large number of small but very sharp teeth and can inflict a painful bite if given half a chance. It hisses like a snake, growls like a carnivore and snaps out at its tormenter. If this fails to deter the enemy, the possum promptly drops dead. It collapses dramatically on to its side, its body limp and lifeless, its back slightly bent and its head turned down, so that it has a half-curled-up appearance. Like the best fakers from the snake world, it displays a slackly open mouth with its tongue lolling out, and its eyes are half-closed.

In this pathetic condition it is completely immune to any pain that may be inflicted upon it. Even if bitten or pecked it continues to lie limply on the ground. What is the mechanism in the animal's nervous system that makes this possible? It used to be a popular theory that death-feigning animals were somehow self-hypnotised, or that they were thrown into such an acute state of shock by their tormentors that they went into an unconscious, cataleptic state, as though from a fit or seizure. Others likened it to fainting – a dead faint that switched off all the animal's faculties. But careful studies in the 1960s revealed that none of these explanations will do. Employing an electroencephalogram to measure the brain waves of opossums when awake, when asleep and when playing dead, investigators were surprised to find that the condition of the brain when death-feigning is taking place is the same as that in the wide-awake animal. There was no sign of any special or unusual wave pattern during the minutes when the animal was apparently lifeless. This study showed that the opossum does not experience some kind of fit or paralytic unconsciousness. Indeed, it must be fully aware of what is happening to it when it is playing dead. In some strange way, however, it is incapable of putting its usual pain reactions into operation. The mystery that remains concerns the almost insoluble problem of whether the possum feels the pain when it is being bitten, but is incapable of responding to it, or whether it feels nothing, even though it 'knows' it is being bitten. In other words, is

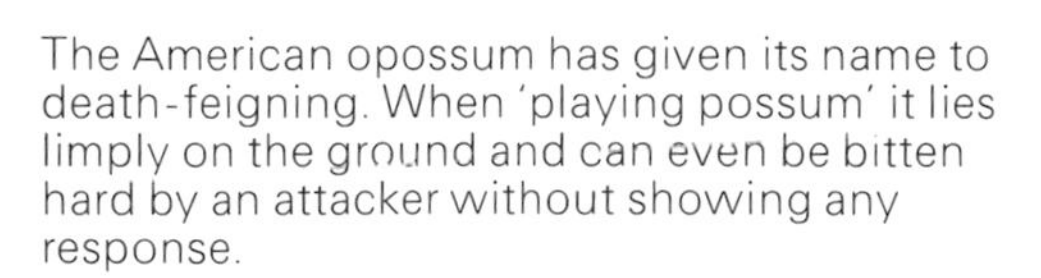

The American opossum has given its name to death-feigning. When 'playing possum' it lies limply on the ground and can even be bitten hard by an attacker without showing any response.

it a sensitive but totally incapacitated animal, or one that is effectively self-anaesthetised? Whatever the answer may be, it is certain that we are dealing with an extremely strange reaction here, and one that might repay much greater attention.

It has been argued that death-feigning in natural circumstances is of comparatively little use and far too risky to become a major defence strategy. If a killer is faced with a quiet prey rather than one that is resisting violently, it will simply be able to enjoy an easier meal. If the prey obligingly offers itself as an immobilised piece of meat, what advantage does it gain? Clearly this view is mistaken or the death-feigning reaction would never have evolved. Despite the superficial stupidity of the action, it must have some special advantage in at least enough cases for it to have been the subject of natural selection.

The answer seems to lie in certain predators' need for prey-movement during the hunting process. Predators that never scavenge and always insist on eating freshly-killed prey will have their appetite switched off by a limp, static object. Many animal killers like their prey to writhe and scream, not because they are sadistic but simply because that is the way they can ensure that they are about to eat fresh, healthy food. For them the seemingly dead animal is far less enticing. If this makes them pause, even briefly, it may be just enough for the prey to escape.

There is another function of playing dead that has been depicted for centuries but which, until recently, was thought to be no more than a fictitious folk-tale. As long ago as the twelfth century there appeared in the early bestiaries an illustration of a cunning fox feigning death, surrounded by birds. He lies upside-down with his eyes closed, his mouth open and his tongue hanging out. The birds are hungry and gather round the corpse to eat it. Then, at the last moment, the fox leaps up and catches them, devouring them instead. Nobody took this fable seriously until in 1961 a Russian film-maker managed to record it actually happening. There in the film is a fox lying limply on the ground. A carrion crow approaches it and prepares to start feasting on its body. All at once the fox leaps round, catches and kills the crow and makes a meal of it. Sometimes nature is just as strange as ancient legends.

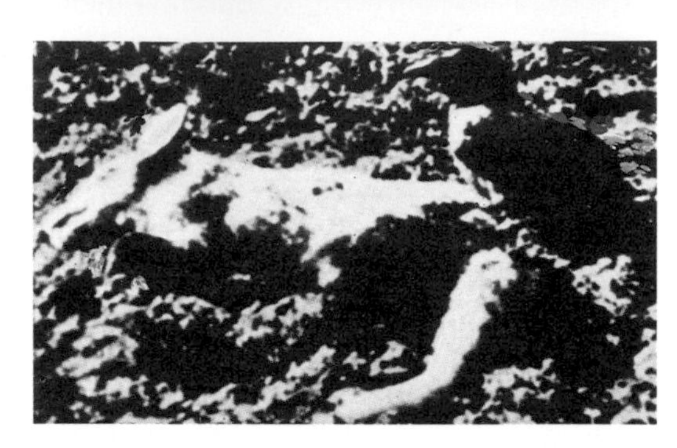

The legend of the fox that traps birds by pretending to be dead has been depicted for centuries. It was shown in the fourteenth-century Queen Mary Psalter and many times in the centuries that followed, but it was later rejected as an imaginative fiction. Then in 1961 a Russian film-maker managed to record the event and prove, for once, that an old legend is also a scientific fact.

Self-mutilation

ANYONE WHO HAS LOST a finger in an accident will be painfully aware that the human species is incapable of regrowing lost appendages. This is one of the ways in which we are inferior to the so-called 'lower' forms of animal life. The best we can do is to grow a covering of skin over the stump where the amputation occurred. In this respect other mammals, and birds, share our short-coming. To find the most advanced animals that show regular regeneration of lost organs, we must turn to the reptiles and, in particular, to the lizards. For them, losing their tails as a way of saving their heads is a common event, and is a specialised process known as autotomy, which literally means 'self-cutting'.

Autotomy takes place when a lizard is not quick enough to elude the attentions of a hungry predator. As the killer strikes at the retreating reptile it is the victim's tail that is most likely to be caught in claws or jaws. At the moment of impact, the tail of the lizard breaks away from its body. The lizard is then able to rush away to safety while the predator, momentarily confused by the suddenness of what has happened, concentrates all its efforts on the disembodied tail. The tail wriggles and writhes in its mouth, or in its claws, and its liveliness preoccupies the would-be killer to such an extent that it hardly notices the rapidly escaping form of the rest of the lizard.

The neuromuscular system of the discarded tail is capable of continuing to produce vigorous twisting movements for several minutes after the attack has occurred. This engrossing spectacle fascinates the predator who may eventually devour the tail, although it has little nutritional value. We know that the trick works, because lizards' tails have been found in the stomachs of a wide variety of predators, including carnivorous birds and snakes.

When autotomy was first observed it was thought that the killer simply ripped off the tail by brute force, but closer study revealed that many lizards, especially geckos, have evolved special break-points in their tails – weak spots that fracture with great ease and make the sacrifice a simple matter. For such animals, the lightest touch on the tail is enough to produce automatic contractions that split the tail away from its base. Some species are even capable of shedding their tails without any contact at all. If such a lizard is caught by the body, it is still the tail that snaps off, separating from its base independently of any external pressure.

It might seem obvious that if weak points are going to evolve in the tails of lizards they would do so between the vertebrae. But this is not the case. Strangely, the fracture planes pass through the *middle* of the vertebrae and not in the spaces between them. In a typical lizard, every tail vertebra from the sixth one onwards has a special 'plane of weakness' where a thin layer of cartilage replaces the stronger, bony tissue. At these points there are also constrictions in the blood vessels and nerves, and weaknesses in the connective tissues. So, when the circular muscles of the tail perform their sudden, massive contraction, the whole of the tail section disconnects with ease. Because of the constrictions in the blood vessels – rather like the necks of hour-glasses – there is practically no loss of blood and the blunt end of the tail-stump quickly dries off and hardens.

A lizard without its tail is at a considerable disadvantage. The tail is used in running, swimming, balancing and climbing, not to mention courtship and display. Lizards that have lost their tails by autotomy do not live as long as those that manage to retain them, nor do they breed as efficiently. But they do at least survive, and after a period of time will be able to re-grow their tails. This is a slow and imperfect process, but it does give them back most of their original caudal functions. The only attribute their regrown tail lacks totally is the capacity to be discarded again. This is because it does not develop ordinary vertebrae, but contains instead a

Many lizards, such as the green lizard (above) and the blue-tailed skink (below), are capable of breaking off their tails when they are molested. A dropped tail continues to wriggle, keeping the attention of the predator focused on it while the prey makes its escape. The mutilated lizard will then slowly regrow a new tail, although it is usually rather imperfect.

central rod of cartilage. This provides a rudder that is capable of assisting in balancing and locomotion, but has no break-points for further 'tail-dropping'.

The age of the lizard is important in determining the quality of its second tail. A young lizard can regrow one that is almost as long as the original. An older lizard tends to produce a somewhat abbreviated, stumpy appendage, and sometimes the regeneration goes horribly wrong and the unfortunate animal finds itself growing not one but two or even three new tails at once. In this condition its mobility is severely hampered, but it still has a chance of survival if its environment is benign.

Of the sixteen families of lizards alive today, eleven of them have species that demonstrate tail-sacrifice. Crocodiles are also capable of some slight degree of regeneration after losing a tail-tip, but the process is far less efficient than with lizards. Some snakes lose their tail-tips with comparative ease, under duress, but unlike their four-legged relatives they are incapable of any regrowth.

Certain lizards – some of the skinks and the geckos – sacrifice another part of their body when they are grabbed by a predator. Their trick is to have a very loosely connected skin, so free from the underlying tissue that a sudden gripping pressure can break it away, leaving the killer with only a mouthful of the instantly shed skin. The now almost naked lizard scampers away to safety and quickly regrows a new outer covering. This is highly efficient, but is no more than a slight modification of the normal skin-shedding process – quite an easy evolutionary step for any reptilian species to make, especially in comparison with the 'natural ligature' device in the lizard's tail.

The only other vertebrate animals to show an efficient form of self-

Invertebrates are more efficient at regrowing lost parts. A starfish can regrow all its arms if necessary, even if the mutilation is severe.

Some crabs are capable of throwing off their arms if they are grabbed by a predator. This common shore crab is in the process of regenerating a sacrificed right claw.

mutilation are certain amphibians, especially salamanders. Some of them aid the deception by raising their tails and wagging them at the approaching attacker, while keeping the rest of the body still. This attracts the predator towards the tail and protects the rest of the body. The tail, when bitten, may then be broken off at a weak spot near the base, where there is a slight constriction. But the 'natural ligature' is of a different kind from that of the lizards. Here the plane of weakness is *between* two vertebrae, instead of through the middle of one. After it has been sacrificed, the salamander's tail wriggles about just like the lizard's tail, but it has the additional advantage that it often contains poison glands which the killer finds pungently distasteful. As a result, the predator is far less likely to attack a second individual of this type of prey.

The tactical loss of part of the body and its subsequent regrowth is far more common among the animals without backbones – the invertebrates. Here it can be observed in worms, starfish, crabs and molluscs. Their less advanced bodies appear to be easier to reconstitute than those of the higher animals.

Starfish and brittlefish will readily jettison an arm or two when attacked, and are capable of quickly growing new arms to replace them. Crabs throw off a leg if it is grabbed by a predator, and scuttle away to the safety of a nearby crevice. If they are dragged out again, they will release a second leg, then a third. But they become less and less likely to make further sacrifices, until, when they are literally on their last legs, they cannot be persuaded to give up any more limbs, no matter how hard the killer grabs hold of them. This increasing reluctance makes sense. A crab missing a few legs may have enough running power left to survive, but a crab that is completely legless might as well be dead.

Some crabs employ an aggressive form of autotomy. Instead of waiting for the predator to attack them, they use their powerful claws to turn the tables. They grab hold of the killer's flesh, clamp tight and then cast off the claw. This has been observed when otters have been hunting crabs. The hapless mammal is left with a pair of savage chelae pinching it tight, while the crab makes its escape. The otter has no alternative but to retreat in agony and attempt to prize the claws free. By the time it has done this, the crab is safe. With the next moult it will have regrown its missing claw, although the new one is by no means as impressive as the original.

Bees and wasps will leave their stings in the flesh of their enemies, employing a similarly aggressive form of self-mutilation. Unfortunately for them this act is final. Not only are they incapable of regrowing the stings, but they also cannot live without them. The survival value of this act lies not with the individual but with its hive community.

Perhaps the most remarkable example of self-mutilation that inflicts pain on the attacker is the breaking off of poisonous papillae by various kinds of sea slugs. The colourful animals – naked, marine snails – often display clusters of finger-shaped bladders along their upper surfaces. When approached by a predatory fish, these little creatures contract their bodies and expand their papillae which they then wave about in the water. The fish snaps at these enticingly offered morsels and finds that it is comparatively easy to bite one off. It discovers its mistake a fraction of a second later, for these strange organs contain poison glands and sometimes also additional stinging-cells acquired from the anemones on which the sea slugs themselves feed. Fish quickly learn to give these conspicuous little molluscs a wide berth, so that the loss of papillae is not too frequent. But even when it does occur it does not provide too much of a problem, for the sea slugs are capable of quickly regenerating them ready for the next naïve fish to attack.

Some sea slugs have conspicuous, brightly coloured growths on the tops of their bodies. These coloured papillae can be enlarged when a predator approaches and are then waved about invitingly. If the attacker bites one off it can soon be regrown. The papillae are highly distasteful and deter fish from making a second assault.

Distraction Displays

THE AMAZING LENGTHS to which parent animals will go to protect their young has fascinated animal-watchers for centuries. As far back as the fourth century, one of the founding fathers of the Greek Orthodox Church, Saint Basil, reported with admiration and astonishment that a parent bird may risk its life for its nestlings by attempting to lure a predator away from the nest. It achieved this, he noted, by pretending to be injured, dragging itself along the ground as if it had a broken wing. In this way it gave the predator hope of an easy catch, something it could not resist. Instead of going closer to the nest and finding the genuinely helpless nestlings, the killer followed the apparently helpless adult bird. At the last minute the parent would abandon its charade and fly off to safety. Saint Basil was so enchanted by this selfless action that he concluded his comments with the words 'If things seen are so lovely, what must things unseen be?'

Certainly, this is not the sort of behaviour that modern judges have in mind when they refer to young thugs as behaving 'like animals'. Seen by romantic eyes it is the epitome of the nobility of nature, but what is the truth of the situation? Are the parent birds really putting themselves at risk for their offspring, laying down their own lives so that their young may live? Even a brief moment of objective analysis reveals that this cannot be so. For if the parent bird were to be attacked and killed by the predator, all its young would inevitably perish for subsequent lack of parental care. So for the trick to work the parent must always be perfectly safe. The risk is only apparent. Nevertheless, distraction displays are, by any standards, remarkable patterns of animal behaviour, and require some explanation.

Displays of this kind are found in a wide variety of birds, but they are most common in the ground-nesting species. There is a special reason for this. Tree-nesting species, if they are being approached by a predator, have little hope of concealing their nests. Having spotted them, the predator will know where they are, even if it is momentarily distracted by the parent bird. The best hope for tree-nests is that they should be inaccessible. Failing that, the parent bird's most useful strategy is to attack the climbing killer as it approaches the nest-site. Diving at it and pecking it may just be enough to unsettle it as it makes the difficult final climb. Alternatively, the parent bird can flee and live to rear another clutch of eggs elsewhere.

The only other course of action for a tree-nester is to hop about in the tree, squawking and screeching noisily in an attempt to break the attacker's concentration. This is a crude form of distraction display, but it cannot be developed much further because of the context in which it occurs. The climbing predator is in no position to start chasing after the cackling parent bird, no matter how tempting it may be. Up in the branches this does not make sense.

Down on the ground matters are very different. For a start, the nest, its eggs, its nestlings and its owners are all highly camouflaged. Ground-nests are so well concealed that predators can easily overlook them even when they are only a few feet away. So it pays the nest-owners to sit tight and hope that the killers will pass them by. If this fails, then clearly the eggs or young are at a serious risk. When the parent bird, with the predator only inches away, finally takes flight, the nest is horribly vulnerable, its position revealed and all too easily accessible. This is the moment at which a major diversion is necessary. To attack the predator while it is in a stable position on the ground is far riskier than to do so when it is clinging to high branches. The diving and pecking routine would have little impact and might expose the nest-owners to serious injury or death. Only powerful birds would stand a chance. For the rest something else is needed, and that is where the distraction display comes into operation. At ground level the nest-owner can make exaggerated movements away from the nest and draw the killer towards itself. The combination of inconspicuous nest, conspicuous display and easy terrain, makes this the ideal context for this type of reaction.

The display must be more than just conspicuous. It must also be tantalisingly attractive to a hungry predator. And nothing appeals to an experienced hunter more than a wounded or somehow disabled prey. If the parent bird can appear to be at some grave physical disadvantage, the killer will find the tempting proximity of the vulnerable adult irresistible and its concentration on the location of the nest-site can be broken. Once it has been led farther and farther away from the nest by the parent bird, it will have difficulty in rediscovering the position of the nest and will, in all probability, set off to search elsewhere if it fails to make a killing.

A successful distraction display therefore requires five stages. First, the bird must sit tight on the nest until the last moment. Second, it must fly swiftly away from the nest and alight some distance from it. Third, it must perform its conspicuous, super-prey routine until it attracts the attention of the predator. Fourth, it must continue this display while moving away from the nest. And fifth, it must be alert enough to judge perfectly the moment for a sudden 'recovery' and flight to safety.

The most remarkable phase of this sequence is the 'super-prey' display. It takes two distinct forms in different species. These have been christened the 'broken-wing' display and the 'rodent-run'. In the broken-wing performance, the realism is so remarkable that even human observers can be fooled when they first encounter it. Having flown away from the nest, the bird flutters to the ground and then pathetically tries to run away, limping and dragging one outstretched wing along the ground. Sometimes it flops to the ground as if exhausted. Its feathers are ruffled, as those of a genuinely sick bird would be, and this ruffling often exposes bright patches of plumage that add to the conspicuousness of the display. If the parent bird is aquatic and is making its display on water, it may turn on its side, flap one wing awkwardly in the air and paddle itself in a circle as if it is hopelessly crippled.

Even while they are staggering about, fluttering and floundering, these mock-disabled birds are fully aware of everything around them. They are not suffering some kind of seizure brought on by the fear of the nearby predator. All their actions are counterfeit. An early observer was amused to notice that certain shore-nesting birds took great care about the way they performed their imitation death throes: 'The end comes slowly, surely, a miserable flurry and scraping, the dying stilt, however, even *in articulo mortis*, contriving to avoid inconvenient stones, upon which decently to expire.'

In other words, these are ritualised displays triggered off by the stimulus of 'predator-near-nest'. Whatever their origin, they are now highly controlled and automatic responses to this situation. They represent an elaborate piece of behavioural mimicry and appear to be inborn. Indeed a Galapagos dove was seen to perform a distraction display on an island where there had been no predators for countless years. When humans arrived, the dove demonstrated a behaviour pattern that it must have inherited from its ancient ancestors and kept in storage ever since.

The 'rodent-run' display takes an entirely different form and is employed especially where the local predatory species are always on the alert for small rats, mice, or lemmings. The bird jumps from its nest at the last minute and runs off making a special squeaking call. Its body is hunched up in a strange way, with its wings drooped, its feathers ruffled and its head lowered. In its humped posture it runs along in a zigzag path, quivering its wings and stopping every so often to make sure that it is being followed. If the predator ignores this tempting performance, the little bird quickly flies back and runs close to it, moving with great agitation. It then starts off on another 'rodent-run', watching to see if it is followed. It

Many ground-nesting birds employ a distraction display that draws an approaching predator away from the nest-site where vulnerable eggs or chicks are situated. The display, seen here performed by a thick-knee (top), a kildeer (above) and a semi-palmated plover (opposite, top and bottom), usually consists of a simulated injury – the body is rolled on one side and an apparently broken wing trailed along the ground. The 'damaged' bird struggles away from the nest, followed closely by the would-be killer. When the predator comes too close, the parent bird suddenly leaps up and flies to safety.

The bravery of a parent bird when defending its nest is remarkable, as is proved by this blacksmith plover's display towards an intruding elephant. Although the elephant is not a predator, it could easily damage the nest, and it must somehow be persuaded to take another route. The little bird is dwarfed by the giant pachyderm, but does not hesitate to perform its distraction display at close quarters.

Even the mighty ostrich has been known to resort to this form of deception. Instead of risking an attack on the predator that is approaching the nest, the parent bird sometimes runs awkwardly away, flapping its wings as if in difficulties. At any moment it can switch to full-speed running. The predator, by the time it has given up the chase, will probably have forgotten the exact location of the nest and will not be able to retrace its steps. The ostrich, on the other hand, will eventually be able to return to the spot it knows so well, where it can resume its parental duties.

continues to do this until the predator is safely in pursuit and is moving farther and farther away from the nest.

There are many minor variations, from species to species, but these two types of distraction display are the basic forms. Their strangeness is a testimony to the sensitivity of the predators. No crude display will suffice to decoy them. The performance has to be convincing, or they will not be lured away from the vulnerable nest or young. And they have to be safe for the performers, so that the nest-owners can survive to continue their parental duties. Their inner mood must remain one of complete alertness to be ready to make the final dash to safety. They may appear to be fainting, collapsing and suffering from shock, but in reality they can be none of these things. Everything must be a sham. In which case, how did they develop these extraordinary actions during the course of evolution? What is their origin?

As with all displays, the unusual movements and postures are the result of a state of conflict. In this case the conflict is between the urge to flee from the predator and the urge to attack it. At the same time there is a conflict between wanting to be *on* the nest and *away* from the nest. Originally this state must have led to a great deal of contradictory and ambivalent behaviour, with the bird being tugged in opposite directions by its opposing moods. Clashing intention movements of coming and going, autonomic disturbances leading to harsh vocalisations and feather erections, and agitated changes of direction could all be expected to occur.

If this were a simple conflict between attack and escape, it would have led to threat displays. These would have evolved in such a way that the displayer would have appeared bigger and more frightening. But intimidation is not the function of distraction displays. Here selection has favoured display developments that make the birds look not fearsome but disabled. The elements selected from the confusion of conflicting movements have this different and very distinctive bias. In particular, a combination in which one part of the animal flees and the other part stays still, is favoured. This has produced the 'broken-wing' display. The flapping wing says 'I am frightened,' and the stationary wing says, 'I will not retreat.' Together, this gives the awkward, asymmetrical display that mimics injury. In the case of the curious 'rodent-run' display, it is as if the bird is running away and hiding at the same time. The strangely crouching posture, with the head down and the body hunched, seems to have been taken from the 'hiding-by-staying-put-on-the-nest' repertoire, while the running away and squealing has been taken from the escape repertoire. Put together, along with a 'run/don't-run' locomotion compromise, they create the rodent mimicry that appeals so much to certain killers.

But whatever their origins, these distraction displays are now reactions in their own right. They have become emancipated from their roots and exist as trigger reactions to the approach of predators. They are, however, finely tuned to the parental condition of the birds. They are most readily shown at the times when the young are hatching and again when they are fledging. These are the most intense parental periods for the birds and the increase in parental arousal affects the frequency of the distraction displays.

Distraction displays are almost entirely confined to birds, but there are a few examples from elsewhere, notably in a fish called the bowfin. This is a nest-building species in which the male guards the young fry after they hatch. They cluster around him in a dense shoal and follow him as he moves about. They are the favourite prey of certain predatory fish and the male parent is reputed to perform a distraction display at certain times. He does this by moving away from the shoal of fry and then thrashing about in the water as if injured. In this way he may succeed in attracting the attention of the predator away from the vulnerable fry.

Other distraction displays, of a completely different nature, are those in which the displayer attempts to distract the attacker from itself. No young are involved. For example, the cuttlefish, when attacked, releases a large blob of ink into the water. Originally it was believed that this was a simple 'smokescreen', a kind of instant fog in which the cuttlefish became hidden and could escape. But careful observation suggests that this is not the case. The blob of ink initially hangs together and does not disperse immediately. Floating in the water it is suddenly more noticeable than the cuttlefish itself. The blob is black in colour while the cuttlefish instantly becomes cryptically pale, and the mollusc sneaks away, almost invisibly, while the attacker assaults the vividly conspicuous blob of ink.

Another self-protection form of distraction is practised by certain orb-web spiders. These animals are frequently attacked by predatory wasps and they divert the killers by wrapping up the remains of insects or egg cocoons and positioning them in their webs. Being roughly the same size and shape as the spiders, they act as decoys, attracting the attention of the wasps and sparing the spiders.

This spotted sea hare (above) is a mollusc that can squirt a cloud of purple ink into the water in response to the approach of a predator. This display distracts the attention of the would-be killer from the retreating form of the prey. The octopus (below) and the cuttlefish both use a similar strategy when under attack, but their ink blob is dark brown or black.

Mobbing

Small birds suffer from a lifelong fear of birds of prey, especially owls. It is an inborn fear that develops when they are only a few months old, regardless of whether they have ever met an owl. In general it results in their avoiding owls at all costs, but occasionally they turn the tables on the killer bird – the tormented become the tormentors. Instead of fleeing they stand their ground and confront the owl. Screeching an alarm call, they attract more and more small birds to the scene until the bird of prey is surrounded by a noisy, angry mob. They now start to harass the bigger bird, calling incessantly and loudly, twisting and jerking their bodies and even making mock attacks. Sometimes a particularly bold individual will risk a real attack, swooping in from behind the owl and striking at its plumage.

This mobbing behaviour never occurs when the predator is actively hunting. It is most likely to occur when the bird of prey is behaving in an odd way. If it is injured or sick it may sit quietly in an unusually visible position during daylight hours. A conspicuous, stationary owl is a major target for mobbing. The small birds gather round and approach it remarkably closely, often as near as 10 feet, and then start to display. The exact movements vary from species to species but in a typical finch such as the chaffinch the body is turned towards the owl, the crown-feathers are raised, the legs are bent, the wings are slightly raised, and the body is jerked quickly from side to side in a crouched, bent-leg posture, while the tail flicks up and down.

If a large predatory bird sits in a conspicuous position it may find itself being mobbed by smaller birds, which call, display, and even attempt to dive-bomb it with pecking attacks. When it flies away it may be pursued until it has left the vicinity – as is happening with this beleaguered golden eagle being chased by a group of aggressive ravens.

Many species have been observed to indulge in this strange mobbing display, including finches, tits, buntings, warblers, blackbirds, thrushes and even little hummingbirds. The hummingbirds become particularly hostile, buzzing around the big bird's head, as close as 2 inches from its face, calling out and jabbing at its eyes. Some of the larger birds, like the blackbirds and thrushes, often risk a little dive-bombing, in which they swoop down on the owl from a distance of about 30 feet, heading straight for it, and then swerve aside only at the very last moment, when they are no more than a foot away. Sometimes they leap at it from behind and claw at its head feathers.

The excitement is contagious, with many new birds arriving and performing the mobbing display without even seeing the bird of prey that is causing the commotion. They witness the behaviour of the other small mobbers and simply follow suit. They become so excited during this gang warfare that humans can approach them much more closely than at other times. And their arousal is so intense that if the owl finally departs they will still go on mobbing for a long while afterwards, as though they cannot calm down to a normal level of activity until some considerable time has passed.

Sometimes a solitary individual is brave enough to act as a one-bird mob and attack a predator on its own, as this roller is doing, worrying a martial eagle.

The beleaguered owl finds the whole encounter highly distasteful and confusing. Its bearing suggests that it is irritated and jarred by what is taking place around it. It becomes increasingly ill at ease until eventually the din and the intrusions become too much for it and it flaps off to find a quieter spot somewhere else. And this, of course, is the function of the combined display onslaught. The owl will not forget the ordeal and may in future avoid that particular neck of the woods. For the local small birds this is a huge advantage.

The actions employed by the displaying birds are clearly those of animals in conflict. Fear of the owl makes the birds want to flee and anger at its obtrusive presence makes them want to attack. In addition there is a considerable element of curiosity on the part of the little birds – they want to learn all they can about this dreaded enemy while it is, for once, sitting quietly.

Although serious research on the subject of mobbing did not begin until around the time of the Second World War, it has been known as a natural

The most common victim of mobbing is the owl, and this phenomenon has been known for centuries. The earliest depiction of it is on a sixth century B.C. black-figure amphora (above) from Greece. This shows that human bird-catchers had already learnt to exploit the mobbing reaction. By placing a tethered owl under a tree, the branches of which had been covered with sticky lime, it was possible to trap small birds as they gathered to mob the bird of prey. Their feet became stuck and they could then easily be caught and killed for the pot.

This seventeenth-century mosaic (above, right) illustrates the way in which different species of small birds may uniquely work together to deal with the shared problem of seeing off a feared predator.

(Below) A mockingbird is here seen boldly attacking a great horned owl.

phenomenon for over 2,500 years. There is an exceptionally beautiful Greek vase depicting mobbing, dating from the sixth century BC. It is a black-figure amphora showing a tree with ten branches and an owl tethered near by. On either side of this scene there crouches a human watcher and the air is filled with small birds – thirty-one of them altogether. Some appear to be stuck to the branches and one or two are distinctly fluttering their wings, as if in some kind of trouble.

What this early artist has illustrated is mobbing behaviour being exploited to trap small birds, which would have been destined to be served up at table. The tethered owl, deliberately employed as a decoy, attracted the flock of small birds to the spot, where many of them alighted on the branches of the nearby tree. And this was their undoing because the surfaces of the branches had earlier been covered with lime to make it impossible for them to take flight again – ever again. The owl was being used as bait.

This reveals a close understanding of the phenomenon and shows that it was well known even at this early date. A little later, in Ancient Roman times, the mobbing display again appears in the art of the day. This time it is on an elaborate mosaic, where an owl is shown sitting in the middle of a ring of attacking animals. Symbolically the owl sits on an Evil Eye that has been pierced by a spear, suggesting that the owl's own 'Evil Eye' has been incapacitated by the mobbing display. Later still, in a Medieval bestiary, three courageous birds are shown in the act of attacking an owl's head. One of these birds is depicted as a magpie. Much later, in the eighteenth century, the Spanish artist Goya shows a caged owl being swooped upon by two free-flying birds, while a small caged bird sits nervously in front of it. The indication here is that the caged finch has been placed close to the owl in order to provoke it into prolonged alarm calling that will attract others of its kind. The netting thrown over the tree in the background is obviously intended to trap these newcomers as they arrive to mob the owl.

In modern times interest has centred around what it is about the owl that makes it the target for this hostile treatment. Field tests using stuffed owls and wooden dummies quickly established which features are important. To be an effective 'owl', a model should possess the following qualities: a big head, a short neck, a short tail, solid contours, brown or

Field experiments with stuffed birds also demonstrate that different species of small birds all close in for the attack on a hated enemy – in this case not a predator but a cuckoo. The stuffed cuckoo is bombarded by a frenzied reed warbler and a belligerent pair of nightingales. The willow warbler is even prepared to assault a stuffed cuckoo head.

grey colouring, a patterned surface with spots or streaks, a beak and frontally directed eyes. The more of these properties a dummy possessed the more strongly it was mobbed, but it seemed to make little difference whether the object was a stuffed owl with real plumage or an owl-shaped piece of painted wood. If the key elements of 'owlness' were absent, or only a few were present, the birds showed some curiosity about the dummy but were not stimulated to perform the full mobbing response.

One particular owl quality that is sufficient by itself to attract mobbing birds is the characteristic hooting sound it makes. In the days when hummingbird feathers were fashionable costume accessories, the plumage hunters of Trinidad used to imitate the hooting of owls to draw the unfortunate hummingbirds towards them and to their deaths.

Although they produce the strongest reactions, owls are not the only animals to provoke mobbing reactions. Other birds of prey may also arouse strong responses if they sit in a conspicuous position during the daylight hours. Also, field experiments with a stuffed cuckoo quickly revealed that this, too, was attacked by small birds. At first the mobbers showed the typical agitated response but then, when the stuffed bird failed to retaliate, they dashed in for violent physical assaults on its body. They savagely plucked feathers from it and were so carried away by their onslaught that they even continued attacking when the stuffed bird was held in the experimenter's hand, close to his face. At one point, to suppress attacks, a large handkerchief was thrown over the stuffed cuckoo, but the enraged birds (in this case, nightingales) would not give up and assaulted the handkerchief instead. Such is the passion of the mob.

Snakes have been mobbed from time to time by groups of ground squirrels. The little rodents make brave, taunting rushes at the deadly reptiles and sometimes even risk delivering a lightning bite to their bodies. Another technique they employ is throwing sand or dust into the snake's face. Put under this pressure and without an isolated adversary on which to concentrate its lethal strike, the snake usually does its best to beat a rapid retreat. Again, safety in numbers saves the day for the prey species.

Some predatory mammals also provoke mobbing reactions. Foxes

Like small birds, small mammals may occasionally indulge in mobbing to rid themselves of a feared killer. Here, in the Kalahari, a group of meerkats is angrily seeing off a deadly yellow cobra.

often avoid the colony sites of breeding seabirds because if they attempt a swift invasion to capture nestlings they are liable to find themselves the object of severe and painful dive-bombing by the parent gulls and their companions. The gulls swoop down, striking at the fox's head, and the predator quickly retreats to safety.

Baboons will gang up on a leopard in a similar way, although this is a risky venture. But here, too, the presence of a group of agitated, hostile prey animals, all assembled together as an opposing force, is usually too much for the predator. Every killer has its own special *modus operandi* – or MO, to use a police term – and it does not take kindly to being forced to attack in circumstances that do not favour its special hunting style. An angry mob creates confusion and prevents the smooth running of its hunting sequence. Its decision is therefore almost always to show restraint and to leave the scene as quietly as possible. The hunt will occur later, on its own terms and in its own time. When mobbing occurs, the situation is far too unpredictable from second to second to suit even the most lethal predator.

Chimpanzees have been observed to indulge in mobbing in certain unusual cases. One intrepid field-worker took an electrically animated, stuffed leopard into the forests of West Africa, placed it in a clearing near a group of chimpanzees, and hid in the bushes near by. When the chimpanzees came close to the leopard, he activated its mechanism, so that it started to move its head. The apes' reaction was astonishing. Film of the event shows that they immediately ganged up on the leopard, rushing at it, screaming, stamping their feet and throwing objects at it. Eventually, one of them attacked it with a large branch, striking it a damaging blow. This is so reminiscent of the way human mobs have all too often treated wild animals they have encountered, that it is just one more reminder of how closely related humans and chimpanzees are.

Food-finding

THERE ARE ABOUT one and a half million different kinds of life forms on the planet today. Of these, 1,124,000 are animals and 359,000 are plants. The animal species are divided as follows (to the nearest thousand):

4,000 Mammals
9,000 Birds
6,000 Reptiles
3,000 Amphibians
20,000 Fish
80,000 Molluscs (shellfish, octopuses and squids)
4,000 Echinoderms (starfish and sea-urchins)
923,000 Arthropods (insects, spiders, scorpions and crustaceans)
9,000 Coelenterates (jellyfish, sea-anemones and corals)
66,000 Lower Forms (worms and microbes)

Every one of these species has evolved its own special way of finding food, but they fall into eleven basic categories. At the top of the scale come the *predators*, animals that live largely by killing other animals. Some are highly specialised and limit themselves to one particular type of prey, refusing all other kinds of meat even though it may be highly nutritious. Others are opportunists, killing anything they can catch and devouring it with relish. The specialists tend to live rather quiet, often secretive lives, while the opportunists scamper hither and thither, always on the lookout for a new kind of meal. Eagles, snakes and cats are typical lazy specialists, and crows, mongooses and dogs are typical busy opportunists. The use of the word 'lazy' here does not imply that those animals are indolent or inefficient but rather that their specialisation is so finely tuned and successful that they are able to spend a great deal of their daily lives sitting, gazing and sleeping. The opportunists, by contrast, must be forever searching and sniffing, exploring and checking.

These are two totally different ways of life, each with its own rewards. The difference becomes crucially important when animals are brought into captivity, as in zoos or menageries. The specialists are then much easier to satisfy – providing they are given the right kind of food, they are content to sit and stare out from their cages, much as they would sit and stare out at their wild landscapes. But the opportunists suffer terribly under these conditions. Even if well fed, they cannot stop searching and exploring, as this has become an end in itself in their behaviour repertoire. Their problem is often that there is nothing to explore except a bare concrete cell, and this frustrates them intensely. To watch such animals biting their own bodies as a way of providing novel stimulation is to see just how inadequate even the most modern zoos are for this type of species. Opportunist feeders must have a rich and complex environment if they are to thrive. Zoos really cannot provide this without giving them enormous, pseudo-natural enclosures.

In the wild, the specialist predators may spend only a very small percentage of their time in the pursuit of food. A kill every few days is often enough for the predators of large prey. After each kill they can feast and rest, then slowly digest the flesh on which they have gorged themselves. This is food of the highest quality and their feeding technique, although always involving the risk of physical injury, is the most efficient of all.

The smaller the prey becomes, the more labour is needed to obtain adequate nutrition. Each victim has to be tracked and killed and yet it provides only a small meal. But its food value is still very high if it is a vertebrate prey, such as a rodent, a bird or a fish. The feeding of insect-eaters is less efficient. They must scoop up or snap up huge numbers of insects each day to fill their stomachs, and each insect contains a large percentage of rather non-nutritious outer-covering. The advantage, however, is that there are plenty of insects about.

The second major category of feeders is the *scavenger* group. These are

The supreme predator is the cheetah (opposite, top) the fastest animal on four legs, capable of reaching speeds of over 50 miles an hour. But its chase is brief, averaging less than 200 yards, and if it fails to catch its prey within this short dash it gives up. If it does catch it, it knocks it to the ground and then grabs its throat in its strong jaws, killing it by suffocation.

African hunting dogs (opposite, bottom) attack their prey in a pack, pursuing them for several miles if necessary. Their secret is stamina rather than speed and when they have worn down their prey the kill is clumsy but efficient. They simply tear at the flesh of the victim until the animal is weak from loss of blood, when they pull it over and start to devour it. Smaller prey are disembowelled and then eaten while still alive.

The lion is an ambush killer. With its heavy body it is only capable of reaching speeds of about 35 miles an hour, some 15 miles an hour slower than many of its prey, and so it requires stealth and surprise rather than speed to catch them. Encirclement of prey is a common strategy, as in wolf hunting, and the frantic victim finds it difficult to escape the group attack. It is knocked to the ground by a heavy blow, seized by the throat or mouth and quickly suffocated. The pride then move in to share the feast.

The raccoon has a strong preference for aquatic feeding. The development of its taste for small water creatures as a main ingredient in its omnivorous diet may be similar to the trend that occurred in our own ancient ancestors during the intermediate stage on the way to becoming the first hunting primates.

the meat-eaters who leave the killing to someone else. The benefit of this feeding lifestyle is that it avoids the dangers of active killing of prey. A carcass cannot injure you. But the food is always of second-best quality – the left-overs from the hunters. And there is always a risk of disease to replace the risk of injury. It must be admitted, however, that vultures always appear remarkably healthy.

At the other end of the scale there are the *particle feeders*, static animals such as burrowing marine worms and other bottom-dwelling aquatic creatures that simply sit and sift the plentiful detritus that sinks to the bed of seas and lakes. They rarely go hungry, but they are easy prey for aquatic hunters, having little or no mobility with which to defend themselves.

Similar to these are the *filter-feeders*, animals that pass a current of water through their bodies and take out of it all the food they need. The largest of the filter-feeders is, however, far from static and is indeed the biggest animal that has ever lived – the blue whale. This vast creature, weighing up to 30 tons, feeds by merely swimming along with its mouth open. It collects its food on a curtain of baleen or whalebone in its mouth and after a single meal may well have transferred to its stomach no less than 2 tons of tiny shrimps, each no more than 3 inches long. The largest of the sharks and rays are also filter-feeders.

Among the birds, flamingos have also evolved strainers inside their mouths and operate as filter-feeders in shallow lakes. They feed with their heads upside-down, sweeping them from side to side. The tongue acts as a pump, helping to suck in small forms of water-life and squeeze them down the throat. Dabbling ducks operate a similar system when sifting through the muddy bottoms of the waters on which they live. With lower forms of life that employ this type of feeding system, such as oysters and other bivalves, the animals are usually stationary and pass the current of water

The diet of this Alaskan brown bear (right) is predominantly succulent vegetation, including roots, fruits and berries, but when the opportunity arises animal foods are consumed with relish. Once a year the up-river migration of the salmon heading for their spawning grounds provides a great feast. Each fish is caught in the water, then carried to a nearby bank and delicately devoured. As a restaurant diner would, the bear separates the flesh from the skeleton, which is left behind after the meal.

(Opposite, top left) River-feeding birds have evolved a highly specialised hunting technique based on great restraint. They have developed the ability to stand stock-still, like statues, until the fish are relaxed and unwary. Then a lightning stab is made, in which the bird has to allow for the refraction of the water surface. Herons are masters of this technique.

(Opposite, top right) The first mammals were small, scurrying, nocturnal creatures that hunted for insects and other small animals. Today's tarsiers retain this ancient feeding pattern.

Birds of prey, such as this tawny owl (opposite, bottom) swooping down on a mouse, are at serious risk in many parts of the world because of pesticides employed to eliminate the pests on which they rely for food. Being highly specialised feeders, there are no alternatives for them and their numbers are dwindling fast.

The world's greatest scavengers are undoubtedly the vultures (above). They detect the presence of carrion by sight. Their excellent long-distance vision spots something suspicious and they fly to it at high speed. Other vultures observe these rapid flying actions in the distance and then move in themselves. In this way a carcass can draw vultures from far away in a short time. Once there they gorge themselves until they are so bloated they can hardly take off.

Starfish (above, right) are the vultures of the seabed. Here a swarm of common starfish are rapidly devouring the carcass of a fish.

Hyenas have always been thought of as scavengers but it is now known that they are also vigorous nocturnal hunters. By day, however, they are more likely to move in on someone else's kill and carry off parts that other jaws cannot crack (below).

through their bodies by beating with thousands of tiny filaments calle cilia. Many of these creatures – peacock worms for instance – look mor like plants than animals.

On dry land the nearest equivalents of the filter-feeders are the *grazers* Like their aquatic counterparts they have to take in a great deal o substance to supply their energy needs. But instead of filtering food fron huge quantities of water they have to extract it from a mass of indigestibl vegetable matter. Grazers have the enormous advantage that there i plenty of their favourite food available, all around them, stretching for a far as the eye can see and regrowing time and again regardless of hov much they crop it. But they have the grave disadvantage that they mus spend almost all their waking hours eating it if they are to obtain a adequate diet. The endlessly repetitive cropping and grinding becomes whole way of life, dominating the daily routine of such animals as rabbit and hares, some rodents, marsupials such as kangaroos and wallabies many ungulates and, among invertebrates, the grasshoppers and locusts.

Seeking more succulent vegetation, many herbivores have turned thei attention to higher sources of food in the shrubs and trees of woodland and forests. Antelopes such as the long-necked gerenuks, the even longer necked giraffes and pachyderms like the elephants and the black rhino have all stretched upwards for their sustenance. Koalas, sloths and leaf eating monkeys are already up there, with a never-ending supply of food a their disposal, but again with the problem of breaking it down an digesting it. For these *browsers*, like grazers, the feeding is easy but tediou putting a greater evolutionary pressure on patience rather than intelli gence. For some reason, however, the pachyderms do not fit this picture both elephants and rhinos exhibit far more complex personalities wit much higher intelligence than might be expected for animals with thei style of feeding.

There is another major category of feeders that concentrates more o the highly nutritious parts of plants – their fruits, nuts, seeds and flower Those that are specialist *fruit-eaters* enjoy a much more luscious form c food, rich in water, carbohydrate, Vitamin C and sometimes oil, but poo in protein. Most fruit-eaters require other kinds of food as supplement and they tend to be generally more omnivorous in their habits. Fruit is s tasty, however, so sweet and juicy, that it is a favourite source of food for

Flamingos are specialised filter-feeders. Standing in shallow water with their heads upside-down they sweep from side to side. This sweeping action is hinted at by the different head posture of these three lesser flamingos feeding at the edge of a lake.

Filter-feeding whales, such as this humpback, have the simplest feeding action of all mammals. They swim along with their mouths open and allow all the small marine creatures floating in the water to collect on the curtain of baleen that hangs inside their jaws. This sieve separates the marine plankton from the sea water and the whale then dislodges the food with its huge tongue and swallows it. The whales also dive to the seabed where they plough through the sediment, filtering all the small creatures living there.

Grazers such as these kangaroos are in a sense the dry land counterparts of the aquatic filter-feeders, but their filtering takes place inside their bodies, where they must extract small quantities of nutritious material from the large bulk of vegetation they take in. The poor quality of their food means that they must devote most of their waking hours to foraging and become little more than eating machines.

huge variety of mammals, birds and reptiles. Together these species assist the plant kingdom in its dispersal and reproduction, via the pips, stones and seeds that are rejected or excreted.

Seed-eaters are better off nutritionally, for seeds are rich in protein and fat as well as carbohydrates and can provide an almost complete diet for many species of birds and rodents. And as food objects they have an additional advantage that they will keep well in storage. Countless millions of rats and mice rely for their survival on their large seed-hoards, and all around the world vast flocks of small finches thrive on their plunderings of seed-bearing plants of many kinds. Both in numbers and distribution these are without doubt the most successful of all mammals and birds.

Flower-feeders are a specialised group of small animals that act as sexual go-betweens for plants. Flowers are concerned with three things: sex, bribery and advertisement. As the sex organs of plants they must solve the problem of how boy-meets-girl when both are rooted to the ground. They do this by bribing insects, birds, bats, and even small marsupials and rodents, to visit them. The bribe is sweet nectar, energy-giving and delicious to taste, which the animals can only collect by thrusting into the flowers and, in the process, covering themselves in the sexual dust called pollen – the botanical equivalent of sperm. When this is carried from plant to plant it automatically cross-fertilises the flower heads that are visited and even if some of the flower-feeders plunder the protein-rich pollen itself there is still enough spillage to ensure adequate levels of pollination. The plants advertise the presence of their bribes with bright colours and vivid patterns – yellow and white for the short-tongued feeders and red, blue and violet for the long-tongued.

Since sweet nectar tends to attract all-comers, plants have become selective. They only allow in the chosen few who will be good, active pollinators. Other species need not apply. The genetic 'selection committee' operates a number of intriguing restrictions. Some plants include a chemical in their nectar that is poisonous to all but their preferred pollinators. Others have evolved deep tubes that only the very long

Browsers are grazers that have moved up in the world. Switching from the low quality grasses they have turned their attention to the more succulent leaves above them. Efficient browsers include the extraordinary gerenuk, the tall-standing giraffe, and the high-reaching elephant.

tongued or long-beaked can penetrate. Still others have curved tubes to suit long, curved beaks. Another technique is to limit the amount of nectar so severely that it only attracts small pollinators and makes it too unrewarding for bigger feeders that tend to plunder without pollinating. Plants that prefer insect pollinators advertise with ultra-violet displays that other animals, such as mammals and birds, cannot see. And those that prefer bat and moth pollinators open their flowers only at night.

Flower-feeding is not a lazy life. It requires great industry, as is clear from watching bees and wasps, or hummingbirds and sunbirds. Increasing the efficiency of their laborious food-collecting, bees have evolved a complex communication system by which incoming bees can transmit information to others in the hive concerning the direction, distance and richness of a new source of nectar. The bee that has been foraging does a waggle dance, the angle of which indicates the direction of the food in relation to the position of the sun; the amount of waggling shows the distance the food is from the hive; and a special wing-fanning display tells how rich the source is.

Hummingbirds, in their adaptation to flower-feeding, have become extreme in many ways – extremely small, extremely fast and with an incredibly fast metabolism. The smallest weighs less than a tenth of an ounce, which is less than some of the bigger moths. The wings beat between 50 and 80 times a second (compared with 17 times a second for a house sparrow) and the hummingbirds are capable of hovering for four hours non-stop if necessary. This means that they can beat their wings more than a million times without pausing. They have the highest energy output per body weight of any warm-blooded animal, and must consume half their body weight in nectar every day. In a bizarre calculation to compare a hummingbird's work rate with that of a human being it has been claimed that should a man expend energy at the rate of a hovering hummingbird he would soon run out of cooling sweat, his body would then heat up to 750 degrees Fahrenheit, beyond the melting point of lead, and he would burst into flames. To keep up their strength, these little birds require between fifty and sixty meals a day. At night, instead of sleeping

The tiny Australian marsupials called pygmy possums have brush-tipped tongues, used for sipping nectar from flowers, but like hummingbirds they must take insects to supplement this diet.

like ordinary animals, they go into a state of suspended animation – a torpor which reduces their energy output to one twentieth of that of normal sleeping. In this way they can get through the night without dying (literally) for a midnight snack.

With all these extraordinary features, a heart that beats at more than a thousand times a minute and lungs with a breathing rate of 250 times a minute, it is clear that the evolutionary trend that has led the tiny hummingbirds to feed like hovering insects has put some amazing demands on their avian bodies. It has also made them extremely aggressive, for each must jealously guard its food source from rivals. Its life depends on a lavish daily supply of nectar and it cannot afford to be generous about sharing this if it is to survive. The same is true of the Old World equivalent of the hummingbirds – the sunbirds. It has been calculated that one of these, the African golden-winged sunbird, must defend a territory containing 1,600 flowers if it is to thrive. The actual size of the territory is not important so long as it contains this number of food sources, and where the flowers are packed tightly together the space needed by the bird is relatively small.

Another, quite different category of feeders is the invisible *food-burrowers*. These are small animals that penetrate their food supply and live inside it, burrowing into the tissues on which they feed. A glutton's dream, this maggot-in-an-apple lifestyle is compensated for its severely restricted freedom of movement and cramped quarters by its unrivalled security. Living inside one's food may not be particularly adventurous but it is remarkably safe.

The burrowers include leaf-miners, gall-makers, stem-borers, wood-borers and soil-eaters. Many of these cause us great distress. The bark beetles that ravage our woodlands with Dutch elm disease, the woodworm that destroys our furniture, the marine wood-borer called the shipworm that for centuries played havoc with our fleets – all these and others have given us a jaundiced view of this all-too-successful mode of feeding among many lower forms of animal life.

Even more hated is the eleventh and final category of feeder – the *parasite*. Insects of many kinds, thousands of species of worms, and countless other more microscopic creatures have abandoned the independent lifestyle and adapted themselves to a host animal without which they can no longer survive. They feed either on the outside of their hosts, as ectoparasites – such as fleas, lice, ticks and bugs – or they plunge inside to

Flower-feeding hummingbirds (opposite) are capable of hovering for long periods while taking the nectar from the exotic blooms of their tropical American habitat. Their hovering flight is made possible by a figure-of-eight movement of the wings. They are also capable of flying backwards and their forward speed can be as high as 26 miles an hour. Each hummingbird consumes half its body weight every day in sugar, but must supplement this basic diet with insects and spiders to provide the necessary protein.

Many bat species have become flower-feeding specialists. With their long tongues equipped with bristly papillae, they feed on pollen and nectar, and they, too, are important plant pollinators in many tropical regions.

(Right) Perhaps the most extraordinary tongues in the animal world are those of the hawkmoths. In some species the tongue stretches to a length of 11 inches when the insect is hovering and feeding on nectar. When Darwin found an orchid with nectar glands hidden deep inside its flower he predicted that one day a hawkmoth would be discovered with a tongue long enough to reach them. Some years later such a moth was indeed discovered and was given a Latin name that meant 'the predicted hawkmoth'.

Parasitic animals, such as this rabbit flea feeding on the blood in a rabbit's ear, have the advantage of a secure food supply but the disadvantage of being completely dependent upon their hosts. The rabbit flea is so adapted to the life cycle of the rabbit that it has handed over the control of its own breeding cycle to its host. When the rabbit breeds, the changes in the hormones in the blood eaten by the flea trigger the parasite's own breeding. This is timed so that the arrival of the young fleas matches the arrival of their new hosts – the baby rabbits.

Vampires do not suck blood – they lap it up like a cat. Flying stealthily up to a sleeping animal, such as a sow, a vampire bat alights silently near by and then makes a small, painless incision with its razor-sharp incisor teeth. The victim continues to slumber while blood trickles gently from the wound and is greedily licked up. The bat's saliva contains an anti-coagulant that prevents clotting and enables it to feed from one wound for as long as a quarter of an hour, taking in up to 40 per cent of its own body weight in blood.

burrow in the tissues, swirl around the bloodstream or luxuriate in the alimentary canal. Many are no more than irritants, but others are lethal and the bodies of the host animals are waging a constant war against them. Elaborate defence systems have been developed to provide hosts with a natural resistance to these invaders – defence systems that help to keep the larger animals healthy and resilient or, should they become infected, to keep them active even if not in perfect condition. Many wild animals are capable of carrying a considerable load of these parasitic forms without showing too many ill-effects under normal conditions. However, once they encounter unduly stressful conditions their resistance is lowered and they quickly collapse. This is particularly common when wild animals are brought into captivity. The strain of the unnatural environment quickly leads to an explosion of parastic activity and the animal soon dies. This was the case with many so-called 'difficult' species before adequate veterinary screening was introduced into the most advanced zoos.

The history of human medicine is essentially the story of our attempt to prevent our bodies being used as a food source by small intruders, and we have been so successful in this direction that we now have a new problem to face in the twenty-first century – how to find enough food sources for ourselves. We are, happily, not food specialists but broad opportunists, ready to devour almost anything. And with present population trends, that is just as well.

Luring Behaviour

WHEN THEY ARE HUNGRY some animals simply gather their food, others actively hunt for it, others lie in wait for it, and still others try to attract it to them. Setting an ambush with an attractive bait is a highly specialised way of feeding, but for those species that have evolved this technique the living is easy. Indeed, it is surprising that more species have not developed the habit of luring their prey instead of pursuing it.

The most famous lure in the animal world must be the living bait of the angler fish. These bulbous, bloated, heavily camouflaged marine fish, with their huge mouths ready to snap up or suck in any small fish that comes close, possess a quite extraordinary dorsal fin. Like other fish-fins it is made up of a number of fin-rays – the stiffeners that enable the fins to be opened and held erect – but the first of the fin-rays no longer looks anything like a fin structure. It has become completely separated from the rest of the fin and now consists of a long stalk with a colourful blob at its tip. This lure varies from species to species, but usually looks like a small worm. It is made more appealing by a vibration of the 'fishing-rod' stalk that makes it wobble and wriggle in the water. Its bright colour and its frantic movement render it irresistible to small fish swimming in the vicinity.

The angler fish spends most of its time skulking inconspicuously on the seabed. It switches from passive lurking to active luring as soon as it spots a likely victim. As the prey swims near, the fishing-rod is swung forward and jiggled furiously. As a result the dancing lure is positioned just above and in front of the angler's face. This patient predator waits and waits until, at the very last moment, when the prey fish is about to bite at the lure, the killer's great jaws open and, in a movement so fast that it can hardly be seen by the naked eye, it siphons the unsuspecting victim into its cavernous mouth. Its ability to suck in water is so highly developed that it rarely needs to attack its prey with a forward lunge, like other fish. It simply sits and inhales its meal. To prevent the lure being sucked in with the victim, the angler fish has the ability to swing its fishing-rod back, out of the way, as it opens its jaws, the whole procedure being performed in a fraction of a second.

There are more than two hundred species of angler fish known today and some of them are the strangest fish ever seen. These are the deep-sea forms, where the brightly coloured bait is replaced by luminous lures at the ends of the fishing-lines. Some of these lines can be retracted gradually

The angler fish lies in wait for its prey, immobile and camouflaged. When the prey fish swims close the angler vibrates the filament on top of its head so that a small, wormlike lure is waggled conspicuously in the water. When the prey approaches closely, the angler suddenly opens its mouth and swallows the victim in a single gulp. As it does this it simultaneously swings back its lure. Failure to do so might prove painful, with the predator trying to swallow part of itself. There are many species of angler fish, two of which are shown here.

when approached by prey fish, so that the victims are gently drawn closer and closer to the killer's jaws, before the final, lethal gulp.

The ultimate deception is practised by one of the deepest of these deep-sea anglers. Instead of waving a luminous worm at the end of a filament, it lights up a special lure on the roof of its wide open mouth. The little victim swims happily into its gaping jaws to investigate this apparently tasty morsel and in so doing performs the final act of its short life.

This is not the only example of a mouth lure. There is an American freshwater turtle, by the name of Temminck's snapper, which employs a similar device. It lies very still in the bottom of the water with its mouth gaping open. Its whole body is camouflaged and barely noticeable. Only one small part of the reptile is moving and that is the worm-like tip of its red tongue. It flickers this deceitful organ at any small fish that happen to pass its way. They pause, stare, approach and examine it carefully. Again, it is the last thing they ever do, and the jaws snap tight. The turtle has once more lived up to its name.

There are many variations on the angling theme. Sometimes the trap is set for specific victims. The batfish has a small lure that normally rests in a face-cover which projects from its forehead. It is particularly partial to the small crustaceans that scuttle about on the seabed and its lure, instead of being dangled up above the head, is pointed down towards the floor, where it is better positioned to attract the victim's attention.

Other fish predators specialise in herbivorous prey and offer their victims a strictly vegetarian menu, dangling not a dummy worm but a bundle of dummy algae.

These are examples of highly evolved lures, but sometimes such extreme modifications are not necessary. Catfish have sensitive 'whiskers' or barbels with which they search for food on the bottom. In some species

Deep-sea angler fish are among the most bizarre of all species. They display luminous lures in the murky depths of the oceans. In some the lure grows above the mouth (above), but in others it is attached below it (below). A close look at one of these luminous lures (opposite, top) reveals its intricate design complete with hairlike outgrowths.

these barbels look like small worms and they attract little fish towards them when the catfish is resting. Being growths from the edge of the predator's mouth, they bring the victims within easy snapping distance and therefore can act as lures with very little modification.

The voracious and gigantic South American horned frogs employ a similar tactic. Their whole bodies remain motionless except for one tiny area. A finger on one of their front feet starts to twitch. When any small animal comes close, the finger moves more vigorously, as though the huge amphibian were a Disney cartoon frog conducting an unseen orchestra. Fascinated by this, the hapless victim approaches closely and peers at the waggling digit. As soon as the prey is in range, the giant frog snaps it up.

Among the mammals there are two bizarre examples of luring behaviour. The first concerns the small relative of the bushbabies called Bosman's potto. This is a remarkably slow-moving creature. Unable to scamper around the undergrowth like its athletic relations, it relies on other feeding strategies, including luring by odour. On its genitals it has special glands that secrete a strong-smelling substance which, for pottos at least, is highly erotic. It also happens to attract a variety of insects, who come fluttering, creeping and crawling to their doom. An over-sexed potto can quickly become a fat potto as a result of the far-reaching appeal of its glandular secretions.

The second mammalian example is so odd that it is worth mentioning even though it may not be true. It is perhaps no more than an African legend, but it is so outlandish that there is just the possibility that it is based on a genuine observation. It concerns a large species of mongoose that is said to climb up into a shrub that is covered in small red flowers. There, it turns round so that its rump is pointing outwards and then proceeds to evert its anus until it looks exactly like one of the red blooms. Watching carefully over its shoulder, it waits until a large insect is about to land on its 'petals', whereupon it swings round, snaps up its prey, and then resumes its luring position once more. It would be fascinating to know whether this was a freak incident involving a constipated mongoose that was being pestered by insects, or whether it has become a regular technique for luring prey.

Whether the mongoose example is a true case of luring or not, there is one instance of an animal pretending to be a flower that does work efficiently as a trap for unwary prey. The African devil's flower is a kind of mantis that hangs from a branch pretending to be a brightly-coloured flower. Most of its body looks like a green stalk, with its rear legs resembling twigs or stems, but where its front legs are attached, near the animal's head, there are flattened sections. These are brightly coloured and shaped like the petals of a flower, with the head of the mantis in the centre. From this vantage point, the predatory insect simply watches and waits. Before long a butterfly arrives to settle on this appealing bloom and the mantis quickly grabs it with its front feet.

Examples of *aggressive mimicry* of this kind, where the animal concerned is not trying to protect itself but to lure prey to their death, were first studied systematically by the Victorian spider-expert E. G. Peckham, and have since been given the name of Peckhamian Mimicry. Spiders themsleves do, in some cases, indulge in luring behaviour, but it is important to make the distinction here between luring and trapping. Every spider's web is a trap, but it does not lure the prey – it simply intercepts it. To be genuinely alluring a predatory act must involve the active *attraction* of the prey towards the killer. Crab-spiders illustrate the difference clearly. One type is camouflaged and lies in wait for its prey. A second type is camouflaged and waits inside flowers where the prey will come because they are attracted by the sight or scent of the blooms. This type is employing the flowers as lures. A third type takes the process further,

Temminck's snapper, also known as the alligator snapping turtle, is a reptilian angler that lies in wait for its prey, with its mouth open and a pink, wormlike growth on its tongue wiggling invitingly, looking just like a tasty meal to passing fish. Should they enter the great jaws to investigate, they are immediately snapped up and swallowed.

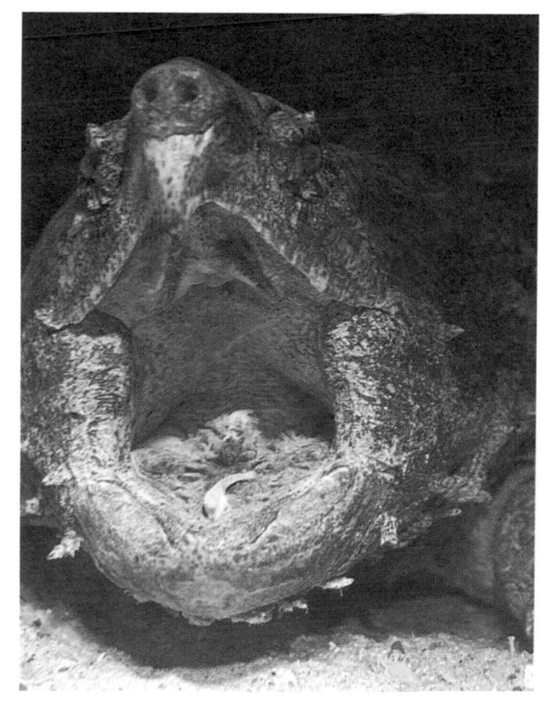

The New Zealand glow-worm, a fungus gnat larva, hangs up threads of glistening beads and illuminates them with the glow from its own body. Insects are attracted to these blobs of light, become stuck to them and are then easy prey.

becoming the lures themselves. These crab-spiders are disguised to look like bird droppings. Their bodies are white with dark streaks and around them they spin a whitish, asymmetrical web that simulates the splatter of the bird dropping as it hits a leaf. Thus concealed, they lie still on top of a green leaf and wait. Butterflies are attracted to them because they seek out the moisture and the nitrogenous salts present in genuine droppings. But as soon as they land near by, the avian excreta leaps into life and demolishes them.

Anyone with a light outside the front door will know how easily it attracts large numbers of insects at night. This same principle is used by certain predators to lure insects to their death. The larva of the New Zealand fungus gnat lives in dark caves where it has been observed to hang up vertical threads covered in sticky droplets and then illuminate them with its bioluminescent body. The light cast from the larva, lying in wait, glints on the droplets, and this attracts small insects which immediately become stuck fast. The larva climbs down, eats its meal, repairs its illuminated fly-paper, and returns to its vantage point to shine again.

Not all cases of luring behaviour have as their end-point the death of their victims. Some prey are less inconvenienced and can be reused at a later date. The little cleaner-fish that provides a marine clinic for parasite-infested larger fish, has a vicious imitator. The true cleaner-fish sets up a special 'grooming-station' to which big fish are attracted whenever they feel the need to have their gills or mouths freed of irritation. They tilt back in the water and let the small cleaner nibble gently over them, ridding them of their small tormentors. The false cleaner has evolved a colour pattern and a style of locomotion that copies very closely the true cleaner, but when the big fish settles in for a pleasant grooming session it is rudely awoken to the fact that this particular barber is not removing parasites but instead is ripping off pieces of skin and flesh. This is luring to snatch a meal, but not to take a life.

Even stranger is the North American fresh-water clam that pretends to have a small fish growing out of its body. This river-dwelling bivalve must somehow transfer its larvae to the gills of a large fish, where they can develop parasitically before dropping off and returning to the riverbed to grow into adult molluscs. This is not easy, especially if the river is fast flowing. The problem is solved by the bivalve having a fish-shaped extension to the edge of its mantle. This is equipped with a false eye and a false tail and undulates gently in the current so that it looks like a small fish. A larger fish, spotting it, moves close to snap it up. As soon as its shadow passes over the mollusc, the dummy prey stops moving and a missile of thousands of tiny mollusc larvae is fired straight at the predatory fish. Before it can retreat it has unwittingly become the godfather of a multitude of parasitic shellfish, which it unavoidably gulps into its mouth, where they attach themselves to the insides of its gills. Again, they do not kill the victim but they do cunningly exploit it.

Finally, there is one form of luring behaviour in which the victim is not even exploited. A small South American fish called the swordtail characin employs a dummy bait as part of its courtship ritual. From each of the male's gill-covers there sprouts a long filament with a bulbous tip. When he is courting a female and becomes sexually aroused his bulbous tips darken and start to twitch. This gives the female the impression that she is seeing a tasty food morsel and she starts to follow it eagerly. The male is capable of raising each filament away from his body and he can direct the one nearest to the female so that its waggling tip is thrust right under her nose. She starts to nibble at it and continues to follow the male as he swims along. In this way he can lure her away and ensure that she stays with him until she is ready to mate.

Young copperhead snakes have a brightly coloured tail. This is waved about as a lure, attracting unwary prey animals, such as this frog, until they are close enough for an easy strike.

Food-preparation

FOR MOST ANIMALS, eating starts as soon as food reaches the mouth. Filling the belly is a serious matter and there is no time to lose. The food object is crushed, chewed or simply gulped straight down. But for some species it is not so easy. Their favoured prey may have an unappetising outer covering that must first be removed before the meal can begin. For them, special techniques of food-preparation become an essential element in the feeding sequence. Indeed, this behaviour is often so important that it develops its own independent motivation. If a captive animal is given its food ready-to-eat, its food-preparation actions may become so frustrated that they will occur anyway, even when not required.

This can be seen in the behaviour of agoutis kept in a zoo enclosure. These long-legged rodents from South America live on the forest floor in their natural habitat, where they dig up roots and other food objects from the earth. These are often covered in a thick layer of dirt and sometimes an inedible skin that they have to remove before they can enjoy the succulent interior. They set about this by sitting up on their haunches and holding the object up to their mouths with their front feet. Working from left to right with their huge incisor teeth, they strip off a section of skin and dirt. Then they rotate the object slightly, like a human diner tackling corn-on-the-cob, and start nibbling again. Strip by strip, they take off the whole outer layer, rotating the food object at the end of every sideways motion. The whole process has about it a highly stylised and rhythmic character, a fastidious mealtime ritual imbued with great concentration and precision. The irony is that should a captive agouti be given a beautifully washed potato it still performs the whole sequence, regardless of the fact that the thin, clean and nutritious potato peel does not warrant it. This is shown by the fact that the captive agouti's eating of the potato has three stages: first it peels the potato, then it eats the potato, then it eats the peelings. If it is given bread rolls, it may even try to peel these, so ingrained is the urge to prepare its food.

Although this pattern of food-cleaning is instinctive for agoutis and occurs even in captive-born ones that have never seen dirty food, it has to be learnt by monkeys. In a troop of Japanese macaques at the Japan Monkey Centre, it was noticed that one young female started to take dirt-covered sweet-potatoes down to the sea to clean them of sand and grit before eating them. This made for a much more pleasant meal and it was soon copied by her playmates. She was one and a half years old when she made her great discovery and the Japanese field-observers watched carefully to see if this invented behaviour would spread through the whole colony, and become a cultural tradition. After a while, the mothers of the young animals began to join them at the water's edge with their own food objects and started to wash these too. Now almost the whole colony except for the mature, adult males were washing their dirty food regularly. The senior males gritted their teeth, literally, and stuck to the old ways.

Eventually the monkey colony saw the birth of new babies and as soon as these youngsters started to feed themselves they all followed the new washing tradition. This applied to both males and females, so when those young males finally became adults the whole colony would be regular food-washers. With the death of the old males, the resistance to washing food would die with them. Here, with this simple pattern of behaviour, we are witnessing at the monkey level the way in which human traditions and cultural differences begin – invented by the playful, exploratory, young members of society, resisted by their elders, but finally adopted as normality by the whole culture.

Perhaps the most famous food-washer in the animal kingdom is the raccoon. In its Latin name *Procyon lotor*, the word *lotor* even means 'washer', so its cleanliness is enshrined for all time. Unfortunately this

The raccoon is well known as a food-washer, but in reality it never attempts to clean its food. In captivity its hunting urges are frustrated, and so it drops its food into its water dish, loses it, searches for it and finds it again. In this way it can satisfy its urge to hunt for aquatic prey, but to the ignorant its actions look like deliberate washing. In the wild, raccoons spend many hours along river banks searching in the shallow water with their sensitive fingers until they locate an attractive food object.

The agouti from South and Central America carefully peels its food before eating it. Usually, as here, the peel is inedible, but if in captivity agoutis are given potatoes they first peel them, then eat the potatoes, and then eat the peel.

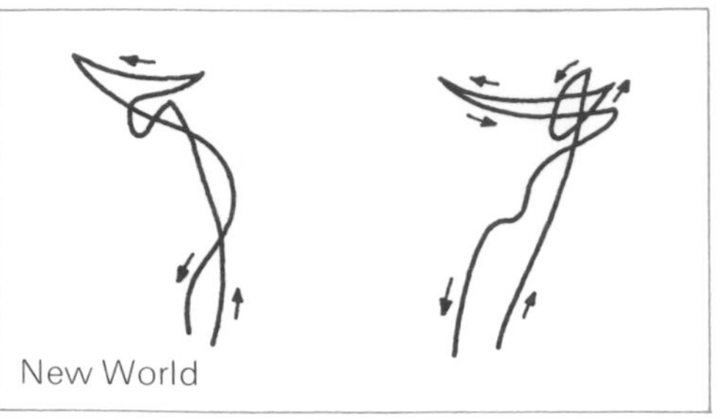

When cats catch birds they pluck the feathers before they eat. Old World cats pluck and shake the head while New World cats pluck first and then shake, as the two curves (above) show.

particular case is a complete myth. Raccoons never wash food – they merely appear to do so. The explanation is simple enough. In the wild, these animals spend a great deal of time at river-banks searching for crayfish, shrimps, molluscs and small fish. They do this by paddling or dabbling with both hands held under the water, their sensitive fingers spread wide and their arms making repeated rotary movements. When these actions lead to the touching of a potential prey, it is immediately scooped up and devoured.

In captivity, raccoons do not have the luxury of a natural river-bank along which to hunt. Their food is given to them neatly in a dish, while their water is placed in another dish near by. Their response to this situation is to take their food over to the water-dish, one object at a time, drop it in the water and then search for it by dabbling. To the ignorant onlooker this had all the appearance of an animal carefully washing its food, but tests proved that this was not the case. Food was washed whether it was clean or dirty, and soft food, such as pieces of bread, was completely ruined and turned into inedible mush by the 'cleaning'. The washing was not a case of food-moistening either, because wet food objects in the food-dish were taken away and dunked more often than dry ones. Observations in the wild revealed that raccoons never carried food objects to water. Dry food was eaten on dry land and wet food was hunted in the streams. In other words, what appears to be food-preparation in this case is in reality 'vacuum hunting'. The captive raccoons have such a strong urge to hunt

for prey in rivers and streams that, prevented from doing this in captivity, they 'invent' hunting by taking dry food to water, losing it, searching eagerly for it and then finding it again. This is so important to them that they are even prepared to destroy soft food in the process, rather than let their frustrated hunting urges go unsatisfied. And they have done this with such regularity over the years that their false washing behaviour has been immortalised in their scientific name.

Another pattern of behaviour that can easily become frustrated in captivity – a genuine food-preparation action this time – is the plucking of prey by cats. In the wild, when they catch and kill a large bird they laboriously pluck out its feathers before settling down to eat it. In zoos, cats are never given dead birds with their feathers on because of the mess they make when preparing them. All meat given to captive cats is prepared as if for human consumption. This is so frustrating for them that when they are occasionally, for observational purposes, given a complete bird, they indulge in an orgy of plucking. One serval cat became almost intoxicated with plucking. After it had removed all the feathers of the pigeon it had been given, it turned to one side and started plucking the long grass next to the dead bird's naked body. It used the same specialised plucking movements on the grass, pulling up the stems and then shaking them from its mouth. It appeared to be completely obsessed with this activity, a vivid testimony to the importance of the special behaviour patterns that are inborn in so many species and which captive conditions so foolishly ignore.

An intriguing sidelight on the plucking actions of cats is that they are performed differently in Old World cats and New World cats. By filming both and then analysing the results in slow-motion, it was possible to reveal that small cats from the Americas, such as the ocelot, pluck the feathers out, raise the head straight up as high as it will go and then shake it vigorously from side to side. Then, the feathers removed from the mouth, the head is lowered straight down again and another mouthful is taken. Old World cats, such as the serval, pluck and shake simultaneously, the lateral shaking movements beginning as soon as the head starts to rise and continuing right to the top of the head lift, and also during the descent back to the corpse. By tracing the movement of the cats' noses on the film, it was possible to draw, frame by frame, 'plucking curves' that show clearly this tiny but significant difference between the behaviour of small cats from the Old and New Worlds. In this way, small behavioural differences can be employed as taxonomic devices for helping with the subtler points of animal classification.

Wherever there is a difficult food object confronting a hungry animal, the animal-watcher can make interesting observations, and there are still many cases that have yet to be studied in detail. Early reports are often less than accurate. Observations on oystercatchers revealed that the technique employed by these birds to open the shells of mussels on the seashore is not a case of the brute-force twisting of the beak in the shell, as had once been suggested. Instead the birds insert their tall, narrow beaks carefully into the live bivalves and make a deft stab at the muscle that holds the shell shut. Before the mollusc can defend itself by closing tight, it is robbed of the ability to do so. The bird can then twist its beak round and prize open the shells without too much trouble, grab hold of the body of the mollusc and, holding it tight, shake off the unwanted shell. This is just one of many such instances where particular species have evolved special methods of dealing with awkward food objects. Further examples are given in the chapter on tool-using.

Japanese macaques in one particular colony learned to take sweet potatoes to the sea and wash them in salt water, removing sand and grit before eating them. This cultural trend in food preparation was started by a $1\frac{1}{2}$-year-old female and spread gradually throughout the colony. After nine years the majority of the group had become potato-washers.

The bill of the oystercatcher, in cross-section, is taller than it is broad and this special shape enables the bird to insert the tip easily into a bivalve mollusc and stab the attachment muscles. The mollusc then cannot keep its protective shell fully closed, and it is a simple matter for the oystercatcher to remove the contents.

Food-storage

PET DOGS often behave strangely with a bone. They can be seen carrying it to the corner of a room, or up on to an armchair, or even into a bed and, once there, making scratching movements with their front feet. After they have, to their satisfaction, completed the digging of an imaginary hole, they solemnly drop the bone into it and then proceed to cover it with non-existent earth. They do this with scooping and pushing movements of their snouts and then finally tap down the invisible earth and quietly depart. The bone may still be lying there in full view, or it may have been successfully thrust under a rag, cushion or pillow. Either way the dog has performed an ancient canine act of food-storage. Although this piece of behaviour has become totally obsolete in the soft world of pet domestication, it is nevertheless carried out in the time-honoured fashion, with all the appropriate elements included in their rigid, primeval sequence.

Dogs do this because they have a food surplus, and that surplus is the essential trigger for all forms of animal food-storage. When wolves in the wild kill a particularly large prey, such as a deer, they cannot gulp down all the meat at one sitting. Each individual wolf is capable of gorging as much as 20 pounds of flesh, but even after the whole group has fed this may leave a considerable amount over. If it is left where it lies it will inevitably be removed by scavengers by the time the wolf pack returns the next day. The only answer is to cache it away, out of sight and out of reach. To do this, the animals tear off large hunks of meat, dig holes in the earth or snow, drop the meat down into the holes and cover them over by pushing the displaced earth or snow back over the top of them. The loose material is then patted down with bunting movements of the snout. When the wolves return on the following day, their bellies less swollen, they dig up the stored food with their front feet, grab it with their powerful jaws, give it a strong

The fox is a scatter-hoarder that buries surplus prey and returns to devour the carcasses at leisure. Here it is examining a food cache at the foot of a tree.

Some wild cats will bury a kill if it is too big to consume at one sitting. This puma is in the act of covering up a large mule-deer carcass to which it may return several times in the days to come.

shake to remove all the dirt clinging to it, and then settle down to consume it. Some individuals prefer to carry their hunks of meat home first and bury them near the den. By doing this they put them in a more protected area where they can keep watch for possible food-thieves.

It is this wolf behaviour that the dog is insisting on performing in the unlikely environment of the sitting-room or bedroom of its adopted home. If it is well fed, all its food is, in a sense, surplus, and therefore suitable for storing. But most of it is soft slop out of a can or a packet and is impossible to carry off to a suitable burial spot. Only a large, solid object will do, and that narrows the dog's options down to the token bone it is so often given by its human owners. Hence its peculiar bone-hiding routine. Some dogs, however, refuse to be outdone by the sogginess of their canned food and can be observed attempting to bury the entire food-dish, slop and all.

Wolves and dogs are not alone in this activity. Many carnivores make caches of food when they have a surplus, hiding it away for later use. Jackals, coyotes, foxes, bears, wolverines, mink, martens and weasels have all been observed hoarding their kills on occasion. Some species only do this when there is an unusually rich supply of prey, and this accounts for what often seems like wanton slaughter in hen-houses. All the killer is doing in such cases is responding to what seems like a splendid opportunity to lay up a food-store. The only problem is that there are so many prey available that the situation quickly gets out of hand and it becomes impossible to carry off and bury all the corpses. The result is that the killer is looked upon as having an 'uncontrollable blood-lust'. In reality it was simply not equipped to deal with a situation where there were literally hundreds of sitting targets. No wild prey would be so vulnerable. In the wild, the killer would be lucky to get even a few extra kills for storing, and so it is not programmed to the unnatural surfeit of a farmer's hen-house.

Some killers, such as jackals, only go to the lengths of food-hoarding when they have the extra burden of cubs to feed. At such times of the year, every ounce of meat is essential and all left-overs are carefully buried and then reclaimed on the following day.

For Arctic foxes, food-storing is a matter of survival. Food is so short in the long winter months that buried larders are vital to their well-being. Conditions can become so severe that it is not surprising that the hoards can become quite impressive. One Arctic fox cache was excavated to find out just how industrious the animal had been when food was plentiful. The animal's natural refrigerator contained: 36 little auks, 4 snow buntings, 2 guillemots and a large number of birds' eggs.

Red foxes also bury eggs, especially when they encounter colonies of seabirds nesting close together. They carry the eggs away one by one, and instead of creating a 'larder' they scatter their hoard, digging individual pits for each egg. In this way, if some other foraging animal manages to find one it will probably not find them all. Foxes keep a good memory of the area in which they bury the eggs, and when they return to excavate them their tracks go straight to the right place without any meandering or hesitation.

Pumas and certain other wild cats occasionally bury food, but they do not dig deep holes like the members of the dog family. Instead they use their front feet to drag debris over the top of the carcass – an action reminiscent of the way domestic cats so often bury their faeces. They frequently pull the carcass into a secluded spot and then heap leaves, twigs and branches over it until it is almost or completely covered. After a few days they return, uncover the kill and feast from it again. Then they cover it once more and continue returning to the site until either they have devoured all the meat or the flesh has become too putrid.

The star of feline food-storage is, however, the leopard. When it makes a

The leopard is capable of carrying a large antelope weighing more than itself high into the fork of a tree, where it is safe from the earthbound predators, such as hyenas. Once stored there, the meal can be eaten at leisure without fear of interruption.

kill the big cat drags the corpse high up into a fork of a tree and wedges it there. It then rests from its exertions and eventually, at its leisure, begins to tear out tufts of fur. Having cleared a patch of flesh, it starts its meal. The strength of leopards in performing their food-storage is impressive, to say the least. They are capable of carrying in their jaws a large antelope weighing much more than themselves, and dragging it up as high as 30 feet. The power of their claws, their leg muscles and in particular their neck muscles is astonishing. They are sometimes so exhausted that it is a full half-hour before they are ready to begin gnawing at the flesh. They cannot complete the meal at one sitting, but stored safely in the air it is protected from the attentions of the earthbound prowlers such as hyenas. For the powerful, solitary leopard it is the perfect feeding system.

A much smaller mammalian killer also makes an impressive food-store, but of an entirely different kind. The humble but fascinating mole consumes large numbers of earthworms and is known to create huge worm-larders that it raids during the cold winter months. Up to a thousand worms have been found in one such larder, and from the prey's point of view this is a particularly gruesome form of storage because they are not killed but merely incapacitated. The mole digs a side-hole in its burrow-system, then hunts down a worm and bites it near its front end. Its bite is venomous and somehow paralyses the worm, which is unable to struggle or escape. The mole continues collecting worms, biting each one in the same manner and adding it to the slimy pile. Eventually, it has enough and seals off the paralysis-pit until the victims, in suspended animation, are needed for a snack. Short-tailed shrews, relatives of the mole, also use their venomous bites to form hoards, but in their case the prey are usually small snails and large beetles.

The mammals that devote most time to the business of food-storage are undoubtedly the rodents. Ninety-one species have been recorded as hoarders and there are sure to be more. The rats, mice and hamsters are famous for their industrious removal of seeds and nuts to their burrows and the squirrels are equally well known for their passion for burying nuts all over their home territory. The reason for the existence of these two different techniques of larder-hoarding and scatter-hoarding is obvious enough. Animals such as rats, with underground burrows, have a suitable location for making a large hoard in a safe place. Animals such as squirrels, living high in the trees, do not.

The origin of larder-hoarding is easy to understand. Rodents that live in burrows are always nervous of predators when they are away from their homes. At the first sign of danger they scamper to safety. If, at the moment of panic, they are in the act of eating, the food object can be retained in the mouth as they flee and eaten later, at leisure, inside the burrow. It is only one step from this to carrying home food repeatedly to make a food-store. Several refinements of this type of hoarding have evolved. Many burrowing rodents have developed cheek pouches in which they can cram large numbers of seeds or nuts. The golden hamster often fills its pouches so full that it almost topples over as it staggers home. Once inside the burrow, it gapes its mouth wide open and removes its collected food with its front paws. The left paw pushes the back of the left pouch while the right paw helps the objects out of the mouth. Then the paws switch roles for the emptying of the right pouch.

The position of the larder varies. In some rodents the food is simply piled up at the back of the sleeping-nest. In others it is placed in a separate side-cavity. In still others, such as the brown rat, there is a special larder-burrow, separate from the main home-burrow. It is usually constructed between the main feeding site and the residential burrow and the rats often visit it and collect a titbit, which they then take home to eat.

The treatment of hoarded food can sometimes be quite specialised. For

Moles make large stores of earthworms. Each worm is paralysed by being bitten at the head end, before being dragged into the underground larder.

Burrowing rodents, such as this little wood mouse, are larder-hoarders, building up huge food stores in or near their underground nests. Some rodent hoards have been found containing over a thousand large nuts.

many species it is simply a matter of hoarding hard foods that will keep, and eating soft foods on the spot. But others go further than this. In the autumn the common vole lays up stores of bulbs for the winter and each bulb is carefully treated before being tucked away, the vole disbudding them as a way of keeping them fresh. The American red squirrel hoards mushrooms, but before they are stored away in special cavities made in a dead tree-stump they are put out to dry on branches. One individual squirrel was observed to collect no fewer than thirteen different species of mushrooms and when they were examined they were all found to be perfectly preserved and fully dried. The same species hoards other foods, such as green pine cones, in damp soil, demonstrating that it has different storage techniques for different objects. The giant kangaroo rat collects seeds and distributes them by scatter-hoarding over a wide area. They are buried about an inch below the surface and remain there until they are dried out, after which they are gathered up and transferred to larders in the animal's burrow system.

The giant kangaroo rat is both a scatter-hoarder and a larder-hoarder, but many species have become specialised purely as scatter-hoarders and make no attempt to collect their stored food into one special place. On the contrary, they are strongly motivated *not* to form larders. The tree-living squirrels and the ground-living (but non-burrowing) agoutis and acouchis scatter-hoard huge numbers of nuts and seeds, trying as hard as possible to avoid burying them where they have already stored previous objects. This activity probably developed from the reaction in these species to the 'food envy' of animals around them. When one foraging animal sees another one eating, it may try to steal the prize. As a result, secret eating may develop, with one animal moving away from others before settling down to opening and devouring a nut. As an extension of this, hiding a nut until a quiet moment arrives is a small step. And hiding it in an unlikely place increases the chances of it remaining undiscovered. From this beginning, it is easy to see how full-scale scatter-hoarding can have evolved, with animals industriously scampering this way and that, digging a fresh hole for every nut they can find. It is significant that scatter-hoarders nearly always camouflage the spot where they have buried their treasure. Squirrels start the hoarding action by digging with their front feet. Then the food is dropped into the small pit and pressed down with the teeth, after which the front feet push the earth back into the hole and pat it down firmly. Finally, the squirrels scrape a few dead leaves or some stones over the spot where they have been working, concealing it as best they can.

Acouchis, which are long-legged relatives of guinea-pigs, use a different sequence. They drop the food into the holes they have dug and then press it down with their feet instead of their teeth. Also, they only use one front foot at a time to fill in the pit, and they do not add camouflage by scraping debris over the area with their feet but pick up leaves in their mouths and drop them on to the spot. These may be trivial differences, but they reveal that the scatter-hoarding pattern has almost certainly evolved independently more than once.

Scatter-hoarding has several advantages. First, the wider the distribution of the food objects, the more difficult it is for raiders to find them. One or two might be unearthed by a hoard-thief, but others will go untouched. Second, if the owner of the buried food changes its home-base it does not have to change the position of its stored food. Third, if, as usually happens, the owner does not rediscover all of its buried nuts, the ones it misses will, by their scattering, develop as a well-planted crop of the animal's favourite food. When the nuts eventually germinate, they will provide a suitable environment for the food-storer's descendants.

Interestingly, some burrowing species are also scatter-hoarders, despite the fact that they have suitable underground retreats in which they could assemble an extensive larder-hoard. The African ground squirrel is one such species and the inference is that it has only recently evolved from tree-dwelling species of squirrels and has not yet readjusted its hoarding behaviour to its subterranean lifestyle.

Unlike the rodents that burrow, those that dwell in trees, such as arboreal squirrels, are scatter-hoarders. They bury large numbers of nuts and seeds over a wide range of their territory. In winter, the concealed food is dug up and eaten (below). The squirrel may not be able to find all its buried nuts, but it can locate enough to make this system extremely important to its winter survival.

Although ground squirrels have underground burrows, some of them scatter-hoard like the tree squirrels, suggesting that they have only recently evolved from arboreal forms. This pallid ground squirrel from Kenya (below, right) has found a site where there is a buried hoard, and is digging for food.

The aquatic life of beavers requires a special type of food-storage. After constructing their lodge, they ram some of their harvested branches into the mud at the bottom of the river. They push them in firmly enough to prevent them being dislodged by the current and swept away. This subaquatic hoard remains untouched during the lush summer and autumn months and is not plundered until the winter snow and ice have arrived. Then, with the river frozen over, they can swim down to the food-store and retrieve branches to provide them with the bark and soft wood they relish. These collecting trips are all done beneath the ice and the beavers are therefore able to bring stored food to their snug dens without any risk from predators or icy winds.

The pika, a short-eared relative of the rabbits and hares, has its own special style of food-storage. In the summer the males start collecting long grasses and piling them into stacks near their den. The grasses dry into hay and by winter can be eaten at leisure. There is a danger, however, that these haystacks, as they become drier, may be blown away in high winds. This is avoided by a remarkable hoarding refinement in which the pikas carry stones to the haystacks and build protective walls with them, shielding the hay from the blast of the storms.

Among birds, food-storage is practised by crows, jays, magpies, woodpeckers, tits and nuthatches. They are nearly all scatter-hoarders and some are as intensely active as rodents. Individual jays, for instance, have been known to bury several thousand acorns, one at a time, in their territories. By contrast, acorn woodpeckers construct something that can be classified as a larder. They select a dead tree and set about drilling small holes in the trunk. They create hundreds of these holes, very close to one another, and then proceed to fill each one with an acorn, wedging it in firmly with their powerful beak.

It has sometimes been rather foolishly argued that animals cannot remember where they have hidden their food. In reality, most species can find their way back to the spots where their buried treasure lies hidden without too much difficulty. In some cases, the memory feats are prodigious, with individual animals recalling not only where they have left their food but also which sites they have already plundered and even which kinds of food have been buried in which locations. If occasionally old hoards are found in such quantity that it appears as though the animals concerned have completely forgotten about them, it should be remembered that many hoarders are favourite prey species and it is much more likely that, in such instances, the hoard's owner has met an untimely end, and has lost its life rather than its memory.

Most birds that store food are scatter-hoarders, but the acorn woodpecker places its acorns so closely together that it can be said to manage a true 'larder'. It drills hundreds of small holes in a dead tree and then rams an acorn into each hole. By keeping the holes together in one special location it is much easier for the bird to remember the site when it becomes hungry during the severe winter months.

Mutual Aid

IN ADDITION TO all the competition and predation that exists between the species, there are a few special relationships that are positively beneficial. In these cases two species live together in a way that is mutually helpful, tolerating a closeness of an unusually intimate kind. Such inter-specific bonds of attachment are referred to as examples of symbiosis, a word that literally means 'together-living'. Because there is mutual aid involved, some authors have coined the term 'mutualism' for this phenomenon. Some examples are easy to understand, but others pose intriguing questions that have yet to be fully answered.

The crocodile bird's partnership with the crocodile is one of the oldest recorded instances of symbiosis, and yet it is still hotly debated even today. Two and a half thousand years ago the Greek historian Herodotus reported that although all other birds and beasts avoided the huge reptile, the little crocodile bird (now identified as the Egyptian plover) was allowed to run all over it and even enter its mouth, where it picked at parasites and ate them. A few centuries later, the Roman author Pliny agreed with this, but felt that the removal of food remnants from around the teeth of the crocodile was the main duty performed by the bird. As time passed, this attractive image of the mutual aid partnership became enshrined in folklore by constant repetition and was widely accepted as a basic truth. The bird gained food from the reptile and the reptile gained the services of a feathered toothpick. The bird not only cleaned the crocodile's mouth and removed any parasites it found clinging there – such as leeches – but also provided an early warning system, raising the alarm in the event of approaching danger. Such was the story handed down from generation to generation.

When Victorian naturalists arrived on the scene to make more detailed field studies, to their dismay they were unable to witness the event. They saw crocodiles basking on the banks of rivers and they saw the little plovers darting about, feeding on insects attracted by the sleeping reptiles, but the famous climax with the small bird standing jauntily inside the giant's open mouth eluded them. Time and again they reported failure until eventually it was agreed that the ancient tale was no more than a legend. This then became the new, accepted truth and was repeated endlessly in ornithological publications.

Recently the strange case of the crocodile bird has been re-examined and new doubts have arisen. Eye-witness accounts of not one but five species of birds cleaning the gums of basking crocodiles have been collected and appear to be authentic. The birds involved are the Egyptian plover, the spur-winged plover, the water dikkop, the common sandpiper, and the grey wagtail. According to one observer, the reptile initiates the cleaning, inviting the bird to enter its mouth by gaping its huge jaws as its small partner approaches.

The only conclusion that can be drawn from this is that, unless certain distinguished field-observers are lying, Herodotus was right after all. But this leaves a puzzle. If the behaviour does occur, why is it so rare that many crocodile-watchers have failed to see it? The answer may be to do with the bird's role as 'watchdog'. If one of its duties in the partnership is to raise the alarm when danger is approaching, it will inevitably react to human watchers as potentially dangerous and become agitated. In this condition, it may still bustle about in the vicinity of the crocodiles but will not be relaxed enough to settle down to full mouth-cleaning. Only very patient watchers, prepared to wait until they become part of the landscape, will be able to catch the birds performing their mouth-entering and teeth-cleaning act. Further field-work is clearly needed.

No such doubts or puzzles surround other examples of symbiosis involving small animals cleaning larger ones. There are many such cases

On coral reefs all over the world, cleaner-shrimps perform a valuable service, removing skin parasites from the bodies of a wide variety of fish, including the large groupers (above) and even the predatory moray eels (below).

Rivalling the cleaner-shrimps, the cleaner-wrasse is a small coral reef fish that not only cleans the outside of the bodies of its 'clients' but is even allowed to enter their mouths and gill covers in search of troublesome parasites. The parrot fish (opposite) is one of its many customers.

fully authenticated today and the 'deal' is always basically the same. The small animal cleans and removes pests and parasites from the larger one, while the larger one refrains from attacking what would otherwise be an easy prey. The small animal gains food and the large one gains hygiene.

This mutual aid system operates all over the world and has thrown up some remarkable business partnerships. In the coral gardens of the tropical oceans, tiny, vividly coloured shrimps perform a cleaning service for many kinds of reef fish. Several species of shrimps are known to act in this way, sitting at their cleaning-stations waving their antennae about as an invitation to cruising fish. Unlike many shrimps they are highly conspicuous, but they are left unharmed. A fish troubled with skin damage or parasites arrives at a cleaning-station and hovers there patiently while the shrimp works its way over its body, sometimes pushing into the gill cavities and even the gaping mouth. Nipping and snipping, it cleans the fish, enjoying a meal of skin debris and parasites as its reward.

A similar relationship exists between certain small cleaner-fish and larger coral reef fish. Like the shrimps, cleaner wrasse are brightly marked with bold stripes of colour, but having no antennae to wave they employ a different 'advertisement', swimming in a dipping, dancing movement that tells bigger fish that the cleaning service is available. Sometimes it is possible to observe an orderly queue of client fish, lined up waiting their turns at a special cleaner-station. When their moment comes they move into position, tilt themselves slightly and gape their mouths as wide as possible, like patients in a dentist's chair. The little cleaner-fish work incredibly hard. One individual was seen to service an incredible three hundred clients in a period of six hours.

The work of the cleaner-fish is crucially important on the coral reef. Without them the bigger fish become pest-ridden and diseased in no time at all. This was proved experimentally by catching all the cleaners on one patch of coral reef. Many of the bigger fish simply deserted the area after this, refusing to continue living in a cleanerless society. The ones that did stay soon became covered in skin infections and were starting to suffer seriously when new young cleaners arrived on the scene and normality returned to the reef. What at first sight appears to be no more than an amusing oddity of behaviour turns out to be a vital part of reef life, with every fish in the community visiting a cleaner-station at least twice a week.

Some client fish become extremely protective of their little cleaners. Huge groupers are serviced by small gobies that put themselves at risk from predators as they go about their cleansing routine. The grouper keeps an eye open for trouble and if it spots danger gives special signals to the cleaners. If a goby is working away inside its great open mouth, it suddenly snaps its jaws together, but halts the movement just before the mouth is fully shut, leaving enough space for the goby to swim out and make for a safe retreat. If another goby is cleaning under a carefully raised gill-cover, the same thing happens. The cover is flicked down but not fully flattened, again leaving room for escape. Once the danger has passed, the grouper will invite further cleaning by staying very still in the water and gaping its jaws as wide as possible. When it has had enough, it shakes itself several times and the gobies leave it in peace and return to their coral outcrop, where they await their next customer.

An even more intimate relationship exists between the remora, or suckerfish, and seagoing giants such as sharks and turtles. The remora has a powerful sucker attachment on the top of its head, a device that has evolved from what was once its dorsal fin and which now gives it a lazy ride over huge distances. It does not harm its host, as was previously thought, but in reality helps it by cleaning its skin from time to time and removing parasites. In return it not only gets food from the cleansing process but can

The remora or suckerfish is sometimes thought of as a parasite, but in reality it is a valuable companion to the large ocean-going host to which it attaches itself, providing a cleaning service for its skin in exchange for food and protection. The sucker on top of its head is a modified dorsal fin.

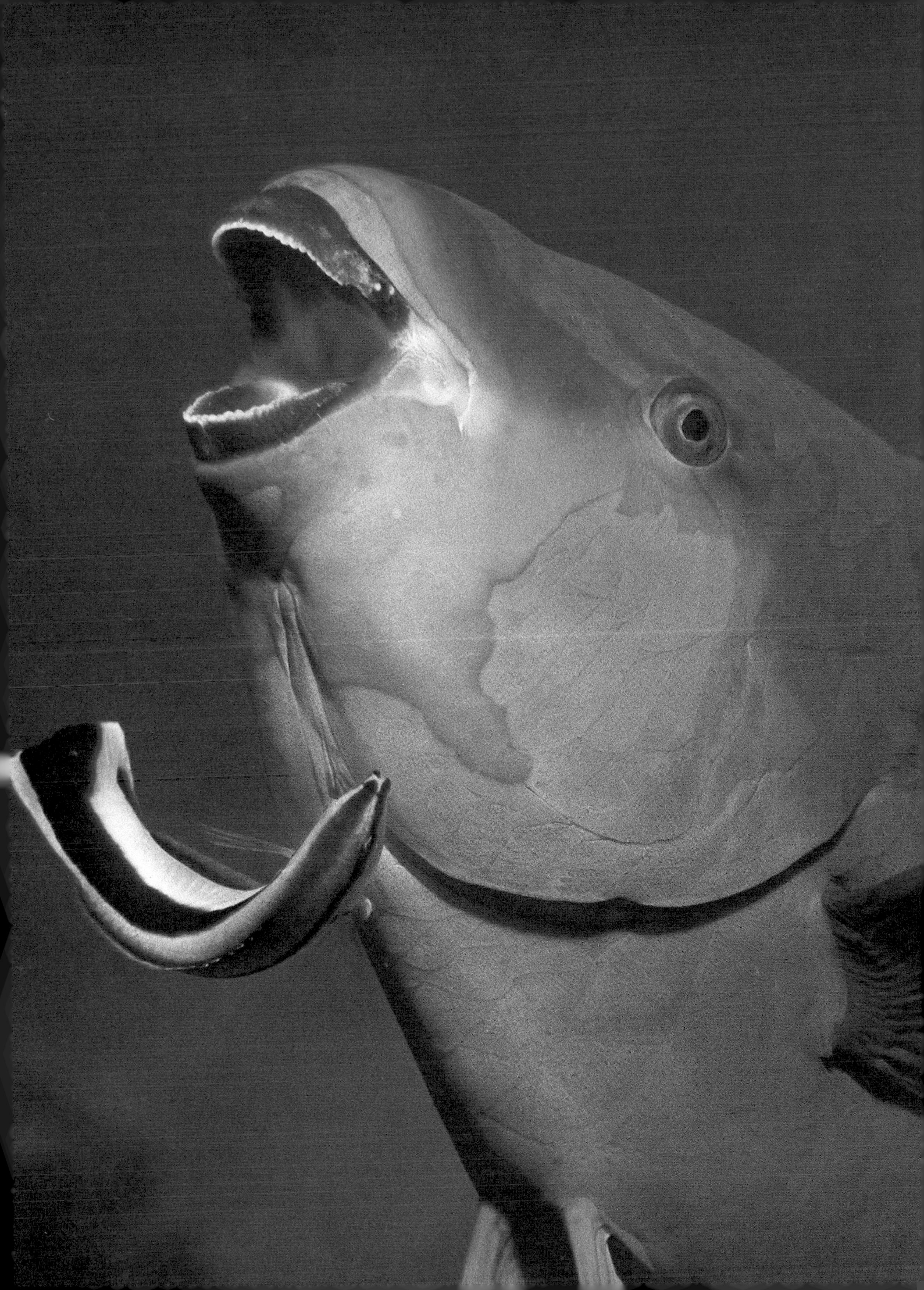

On the African plains the oxpeckers are the local cleaners for the larger hoofed animals. The main task is to remove blood-sucking ticks. This they do with a special scissoring movement, laying the bill sideways on the hide of the host animal's body, even if they are upside-down on its belly. If danger threatens them they quickly run to the blindside of their host. They are common on African buffaloes and even on giant forest hogs, which show a remarkable indifference to their scampering presence.

also join in the messy feasts when the bigger partner is gorging itself, and when there is plenty to spare. In addition the remora gains protection from other predators, which would never dare approach such a monster as its ever-present partner.

Mammalian heavyweights employ cleaner-fish occasionally. Wallowing rhinos have been observed to benefit from small fish that pick off their ticks, and hippos often open their jaws and allow labeo carp to service their gigantic mouths.

Some seabirds have occasionally been observed acting as cleaners. Gulls have been seen to remove parasites from the backs of ocean sunfish, which protrude slightly from the water. Sometimes these enormous fish will roll over on to one side, exposing a greater skin surface to their attendants. Grey phalaropes have even been spotted cleaning the protruding backs of sperm whales as they cruise gently through the open seas.

The most famous bird-cleaners, however, are the oxpeckers that sit on the backs of many species of hoofed animals. A common sight on the plains of Africa, the oxpeckers spend almost their entire lives on their partner's backs. They rest and sleep there, clinging on tightly with their extremely sharp claws. In the breeding season they perform their courtship displays and even copulate there. As a final indignity they pluck out long hairs from the tails or manes of their partners with which to build their nests. Throughout it all, their mammalian aircraft-carriers remain apparently indifferent, prepared to suffer small discomforts in exchange for the invaluable pest-removal. Whether antelopes, giraffes, zebras, cattle, warthogs or rhinos, they badly need the oxpeckers, who gorge on the blood from the bloated ticks and blood-flies that torment the huge bodies of the otherwise helpless mammals.

Cattle egrets perform a similar service, but they are less intimate. Although they do act as body cleaners, they also spend a great deal of time walking on the ground near their partners, enjoying the feast of insects disturbed by the large feet and the grazing mouths. In North America, the cleaner-bird role is taken over by the brown-headed cowbird. Until the coming of man, these birds serviced the vast herds of american bison, but with the slaughter and near-extinction of these native bovines the cleaners have transferred their attentions to imported domestic cattle. Back in the bison glory-days, they had the problem of staying with the migrating herds throughout the year and this meant that they could not pause long enough to nest and incubate their own young. Instead they had to become cuckoo-like in reproduction and lay their eggs in the nests of any birds that

Other cleaner-birds include the white cattle egrets, seen here on water buffaloes in India (below), and occasionally white-naped ravens (bottom) which may take over the duties of the oxpeckers on the backs of the African buffaloes.

happened to be breeding near by. With their newly adopted hosts this is no longer a necessity, but their ancient pattern of breeding behaviour has so far survived.

All these avian mammal-cleaners provide an important, additional service at no extra charge. If they spot a predator before their partners manage to do so, they sound the alarm. And this is quite likely to happen when an experienced predator approaches from down-wind and can be seen more easily than smelled. If the mammalian partners in question are deeply engrossed in tackling a particularly tasty tuft of grass, and ignore the screeching of the small birds on their backs, the birds become frantic, and with good reason, for they *need* their partners, which have virtually become their home territory. If the predator makes a kill they will have to find a new home, and that may not be easy if all the other local mammals are 'occupied' already. So their alarm displays reach panic proportions, with the birds jumping up and down on their partners' backs and finally, in desperation, rushing to the head end and drumming painfully on the unwary mammals' skulls with their beaks. The professional cleaner's life is a demanding one.

Anyone visiting the extraordinary Galapagos Islands will have noticed small brightly coloured crustaceans moving about among the densely packed colonies of marine iguanas. These conspicuous red rock crabs are yet another example of cleaner symbiosis. The iguanas permit them to

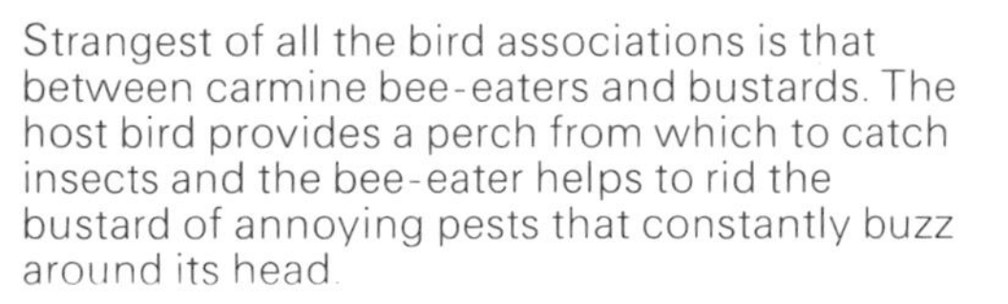

Strangest of all the bird associations is that between carmine bee-eaters and bustards. The host bird provides a perch from which to catch insects and the bee-eater helps to rid the bustard of annoying pests that constantly buzz around its head.

undertake quite painful removals of skin-ticks, sometimes adjusting their bodies carefully in such a way that the crabs can get a better hold on the pests, and the crabs, as with all cleaners, benefit from the titbits they obtain in this way.

Although the cleaner-feeder relationship is the most common form of mutual aid, there are others that are more complex and more intriguing. The sea anemones that festoon the coral reefs of the world carry millions of tiny stinging-cells that are capable of immobilising ordinary fish, which are then drawn into the anemone's mouth and devoured. But there is a group of about thirty species of small reef fish all of which are immune to the poison in the sea anemone's tentacles. Known as anemone fish, damsel fish, or clown fish, these brightly marked reef-dwellers spend most of their lives lolling in the lethal tentacles of their poisonous partners. Their secret defence is a thick layer of special mucus that completely covers their bodies. It possesses a chemical quality that stops the anemone's poison-cells from firing when they are touched by the fish. The chemical involved is still unknown and the precise way it operates remains to be explained, but it somehow manages to deceive the anemone into thinking the fish is not food. Some investigators believe that the fish's trick is to transfer part of the anemone's mucus covering to its own body-mucus. The anemone's many tentacles are constantly rubbing against one another, and must have some kind of sting-inhibitor covering them or the animal would keep on stinging itself. If the fish thoroughly impregnates its own mucus shield with this inhibitor, it will give the impression that its body is part of the anemone and it will be completely safe. This is indeed what seems to happen. A special characteristic of these particular fish must be the possession of a body-surface mucus that will 'marry' with the tentacle mucus. Other fish lack this property, the mucus marriage never takes place, and they are stung to death.

The immunity of anemone fish may have a genetic basis but it requires considerable acclimatisation to function properly. The fish must be in prolonged contact with their particular partners every day for it to continue to work. If they had to switch from one type of anemone to another they would have to start acclimatising all over again. And if they were removed from their own anemone for six weeks, on their return to it they would have lost their special immunity. Again they would have to begin afresh, recharging their mucus with its chemical defence. The fish's method for doing this is to make a series of increasingly intimate approaches, beginning with a very gentle, fleeting contact and then gradually increasing both the strength and the duration of its touches, repeatedly stroking the stinging tentacles with its body as if chemically calming them, little by little. The anemone fish lays its eggs on a rocky surface almost in contact with its anemone. From this vantage point, the young fish are able to start developing their immunity at a very early age, as soon as they hatch from the eggs.

The advantage the anemone fish gains from its relationship is obvious enough. Protected by the tentacles, it is safe from any predator. The advantages the anemone gains are several. First, its resident fish will act as a lure to other fish. They approach to kill and eat the anemone fish, only to be stung to death and consumed by the anemone. Second, certain reef creatures that are capable of devouring the anemone's tentacles, stinging cells and all, are attacked and driven away by the protective anemone fish. Third, when the fish itself is eating, it will inevitably drop particles of food and these can then be consumed by the anemone. And fourth, the anemone fish acts as a cleaner, removing debris and damaged or infected tissues and improving its partner's general condition and health.

Other reef fish enjoy a similar relationship with sea urchins, but in this case they do not have chemical weapons to neutralise. Instead they must

The marine iguanas (above) of the Galapagos Islands suffer badly from skin-ticks, but these are dealt with by the small red rock crabs that live unmolested in the midst of the lizard colonies. The anemone fish (opposite) live among the stinging tentacles of sea anemones, where they enjoy protection from other, predatory fish. They avoid being stung to death by coating their bodies with a protective mucus. Certain crustaceans (below) also set up a commensal relationship with large anemones.

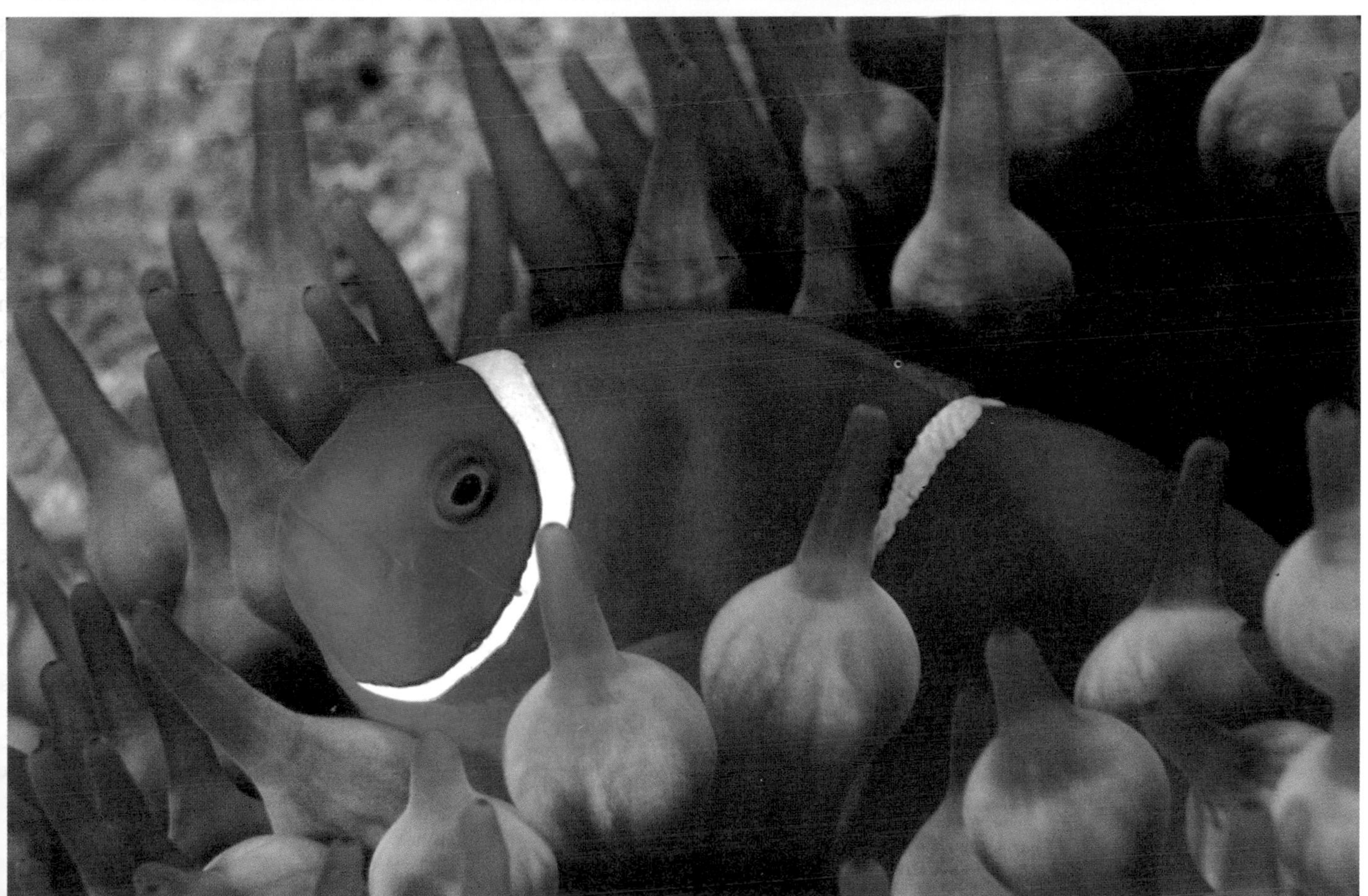

avoid being skewered on the urchin's amazingly sharp spines. Two kinds of fish that have specialised in urchin-cleaning, the shrimp fish and the cling fish, have developed slender bodies and acrobatic swimming techniques, enabling them to move about freely inside the spine-zone, where predators cannot follow them without painful consequences.

Another reef creature that employs a defensive partnership is a tropical crab that carries a small anemone in each claw, like an old-fashioned gun-fighter. When approached by a predatory fish, it thrusts its stinging anemones into the attacker's face. The crab's claws have evolved into specialised anemone-holders and it has to use its front legs to perform the feeding actions that would normally be carried out by the claws.

Other crabs employ anemones as a kind of living camouflage. Certain hermit crabs make themselves inconspicuous by wearing anemones on top of the snail shells in which they live. If they are nevertheless attacked, the stinging cells in their partner's tentacles will be there to defend them. Even the octopus, which relishes small crabs, has been seen retreating rapidly from this particular partnership.

When the hermit crab moves into a new shell, its partner has a problem of house-moving. In some species the anemone simply leans over, clings on to the new shell with its tentacles, releases its foot from the old shell, and swings itself across to its new home. In others, the crab itself makes the transfer.

All these crab-carried anemones benefit from the partnership because of the movements of the crabs and their prolonged feeding activities. The movements keep taking the anemones to new locations, making their diet more varied, and the messy feeding of the crabs provides welcome titbits.

Yet another crab, the sponge crab, has a similar relationship with a red sponge. It snips off a piece of sponge and holds it on top of its shell like a small orange hat, keeping it in place with special hooks on its rear legs. Soon the sponge starts to grow in its new location until it covers the whole upper surface of the crab, providing it with a living camouflage. As an additional benefit, this is a camouflage that tastes bad and will therefore repel even those predators that see through the concealment device. As with the anemones, the sponge benefits from being transported to new feeding grounds as the crab moves about.

Sponge crabs denied sponges to wear on their shells become acutely distressed and will eventually use any substitute they can find. In captivity they have been seen, rather pathetically, cutting up small pieces of cardboard or paper and tenaciously holding these on top of their shells, patiently waiting for them to grow. Like so many symbiotic relationships, this is an extremely intense one and has become an essential part of the animal's behaviour repertoire.

Perhaps the most fascinating example of all these mutual aid systems is the one that exists between a little fish called Luther's goby and the blind shrimp. They live together in a small burrow in the sand in regions where there are no suitable rock cavities in which the goby can hide. The shrimp's contribution to the partnership is to make the burrows, a task at which it excels. The goby's contribution is to guide the blind shrimp on feeding trips away from the burrow. It does this by swimming in tandem with the shrimp. The little crustacean follows by keeping its antennae in contact with the goby's tail. If there is any danger, the fish gives a signal by wagging its tail and the pair retreat to their burrow together. This must surely be one of the strangest partnerships in the animal world.

A particularly elegant relationship has developed between the bitterling fish and the swan mussel. This freshwater species of fish lays its eggs inside the mussel's body, the female depositing them there via a long tube attached to the underside of her body. The male then ejects his sperm which the mussel automatically 'inhales' and the fertilisation takes place

(Top) Some hermit crabs carry anemones on their backs. The crabs have the advantage of camouflage and also protection from predators, while the anemones can enjoy a richer and more varied environment.

(Above) Sponge crabs cover themselves with living sponges. These provide camouflage for the crustaceans and at the same time the sponges benefit from the greater variation in their environment and therefore a richer source of food. In this example the whole of the top of the crab has become immersed in orange sponge.

(Below) An intimate relationship may develop between a shrimp and a goby. They share a burrow on the sandy bottom. The shrimp digs the tunnel and the goby protects the shrimp.

within this highly unusual, living nest. The young fish hatch and develop inside the mollusc and eventually swim to freedom. On this basis there does not appear to be a mutual advantage in the relationship. How does the mussel benefit? The answer is that its own eggs hatch just as the fry are about to leave its body and disperse, and the mussel larvae attach themselves to the little fish. After a while the larvae drop off and in this way the mollusc can scatter its young all over the stretch of water where it lives.

Symbiotic relations between different mammal species are rare, and not particularly intense. On the plains of Africa there are many mixed herds, with zebras, wildebeest and various antelopes mixing freely together. This rather casual symbiosis has the advantage that the different animals can pool their defensive alertness. One species may have better vision, another better hearing, and yet another better scent-detection. Together they have a greater protection against attack than if they move about separately. And in most cases their food specialisations mean that their grazing is not too competitive.

A similar relationship exists between certain antelopes, such as impalas, and troops of baboons. The antelopes can act as sentinels and the baboons as fierce defenders. But neither species is strongly dependent on the other and both occur separately more than they do together. So, compared with the marine examples of symbiosis, this is a very mild arrangement.

One really strange land partnership involving a mammal is the case of the ratel, or honey badger. This African and Asiatic carnivore, related to the weasels, is a solitary animal that has a passion for honey. It is assisted in its quest for bees' nests by a small bird called the honeyguide. When this bird spots a hunting ratel, it chatters loudly and jumps about in an agitated display. The ratel replies with a peculiar hissing chuckle and the two set off together, the bird keeping about 15 to 20 feet ahead of its partner. The bird leads the ratel along, keeping up its noise until it comes to a bees' nest, when it falls silent and waits. The ratel attacks the nest, tears it open and eats the honey. When it has finished, the bird moves in and devours the grubs and in particular, the wax. In this way they share the spoils. In parts of Africa, the local humans have replaced the ratel in this particular symbiotic relationship and the bird has proved invaluable in helping with the search for nutritious honey.

There has been some argument about the exact sequence involved in the partnership between the honeyguide and the ratel. One possibility is that the bird first finds a bees' nest, and then sets off in search of a ratel, finds one, and guides it back to the specific nest it has discovered. If this is what happens, it suggests a highly structured behaviour sequence, and incidentally one that cannot have been learned from the bird's parents, since honeyguides are parasitic breeders like cuckoos and never know their true parents.

The alternative is that the bird is programmed to behave in an excited manner whenever it encounters a hunting ratel intruding into its territory. By displaying and calling and moving slowly away from the earthbound carnivore, it attracts it to follow and leads it deeper and deeper into the forest. It does this without taking any specific direction and simply continues to advance until eventually it inevitably comes across a bees's nest, when it falls silent. The ratel stops too, and then locates the nest and tears it open, fulfulling its part of the mutual aid system. In this sequence the bird's behaviour is less organised and it is this version that is favoured by many observers today, rather than the more planned act of guidance. But we need more evidence to be certain.

The bitterling deposits its eggs inside the swan mussel. The mussel provides a safe nest for the young bitterlings and the bitterling fry provide a dispersal mechanism for the mussel larvae.

Herds of impalas and troops of baboons on the African plains often move about together, to their mutual advantage. The baboons are protected by the alertness of the antelopes to danger signals and the antelopes benefit from the defensive ferocity of the baboons should there be an attack.

Drinking Behaviour

LIVING IN A WORLD of bottles and barrels, taps and tanks, the problem of thirst no longer looms large for us. Yet for many animals finding something to drink is one of the basic challenges of life. For some it is a daily task involving long journeys to a suitable watering-hole. For others it is a constant struggle against conditions that are either too dry or too salty to provide a plentiful supply of liquid nourishment. In extreme cases the problem is so difficult that the animals have given up drinking altogether, obtaining their liquid in other ways.

The well known koala from Australia is one of the non-drinkers. Its very name, taken from the Aboriginal tongue, means 'no drink'. A tree-dwelling food specialist adapted to surviving solely on eucalyptus leaves, the koala is horribly clumsy when moving about on the ground. It is so vulnerable there that it seldom ventures out of the branches unless it is driven down by the need to move to a new food tree, or to eat a little gravel as an aid to digestion. Only in the very hottest weather does it occasionally descend in search of drinking water. For most of its life it simply goes without, obtaining all its liquid from its food. There is moisture on the outside of the leaves and juice inside them, and on these sources it must rely for all its normal water requirements. On this appalling low-quality diet, it is not surprising that the koala is one of the least lively of mammals. It is lethargic enough to make a sloth look like a champion gymnast and it spends eighteen out of every twenty-four hours sleeping in the fork of a tree. This is presumably the only way it can avoid working up a good thirst that it cannot quench.

The koala hardly ever drinks, taking all its water from the leaves it eats. Several acres of trees are needed to produce food for one group of koalas, each adult eating between 2 and 3 pounds of leaves a day.

Other non-drinkers seem to have dealt with the problem rather more efficiently and somehow manage to maintain a more active lifestyle even in the total absence of standing water. The little kangaroo rats living in the very driest of deserts, must often live out their entire lives without ever taking a single drink. They avoid thirst by moistening their food. They do this by carrying the air-dried seeds they find above ground down into their burrows, where they store them in the cool subterranean tunnels. There, the seeds attract moisture and absorb water vapour from the earth before they are eaten. In this way these small rodents, like the koalas, can obtain their necessary liquids from their food, but unlike the large marsupials they are able to remain highly mobile and fast moving. How do they manage it?

Careful observations of kangaroo rats, both in the wild and in captivity, have revealed that these extraordinary rodents have several special adaptions to life in a waterless world. First, they assiduously avoid moving about above ground during the daytime, when overheating would demand the use of body water for temperature regulation. They only emerge from their deep burrows during the cool of the desert night. Second, they have no sweat glands and lose virtually no water at all by evaporation through the skin. Third, they have managed to reduce water loss from the lungs to a minimum during normal breathing. They achieve this by keeping the temperature of the expired air unusually low. Fourth, they produce highly concentrated urine in which salt is at twice the level found in sea water. Finally, the kangaroo rat's faeces are extremely dry and the rate of water loss from them is only one-fifth of that from the faeces of ordinary white rats.

The most surprising finding of these kangaroo rat studies was that these small desert rodents are so well adjusted to their arid lifestyle that if given juicy green foods they ignore them and continue to eat their usual dry seeds. This is a striking example of the extremes to which an animal's physiology can be driven by harsh environmental conditions. But theirs is not the only solution to the problem of living in hot deserts. The American pack rat has adopted a different strategy. Instead of reducing its water intake it has managed to develop a tolerance for a type of water that is poisonous to other animals. It lives in cactus deserts and although it rarely gets the chance to drink free-standing water it manages to obtain all it wants from eating the insides of the highly succulent cacti. Though these plants are 90 per cent water, their flesh also contains oxalic acid – a substance that in large quantities is usually highly toxic. However, they provide a vast supply of liquid for the pack rats which can metabolise the acid.

The Arabian camel adopts yet another strategy. It does not store water in its hump, its stomach, or any other particular region of its body, as was popularly believed. It drinks greedily when water is available, consuming as much as 30 per cent of its body weight in a matter of minutes. One large male was seen to gulp down 22 gallons. But it then uses this water very sparingly during its long desert journeys, on which it can travel for as much as seventeen days before its next drink.

The camel's water conservation is based on a remarkable ability. Any human being who has suffered from a fever will know all too well what a high temperature can do. As the thermometer mounts to 101 or 102 degrees Fahrenheit, discomfort climbs to agony and hallucination. We are simply not capable of tolerating more than a very small fluctuation in body temperature. The camel has overcome this problem. During the intense heat of the day its body can heat up to 105 degrees before the animal even begins to sweat and at night its body temperature can drop to a mere 95 degrees without causing ill effect. By allowing the cold night air to chill it, the camel then gets rid of the excessive body heat, without the usual water loss that has to be employed by other animals as a cooling device, and starts out the new day ready for gradual re-heating.

This is not all. The camel also has the ability to bypass its kidneys. Most animals must lose water in their urine because they have to wash away dangerous waste products. But the camel is capable under certain circumstances of re-routing its waste products from its liver and resynthesising them as proteins in its stomach. In a further refinement to deal with extremes of drying out, the camel is also capable of preventing its blood system losing water at the same rate as the rest of the body. Loss of water from the blood plasma causes the greatest damage in other species during dehydration, and it is the camel's special ability to avoid this that permits it to make its marathon journeys across the burning sands of the Middle East.

Small animals suffer just as much as large ones in extreme desert conditions and, at the other end of the scale, the little darkling beetles have to tackle the problem in their own special way. On the shifting dunes of the great Namib desert in south-west Africa, these beetles have only one way of obtaining water. There is less than half an inch of rainfall in an entire year and the only source of water is a light fog that appears at dawn and briefly dampens the sand before the heat of another day takes over. At night the beetles bury themselves in the sand and gradually lose their body heat. When they emerge at dawn they are cold enough for the dew to condense on their backs. To convert this into drinkable water, they crawl to the crest of a dune and stand still with their rear ends held higher than their heads. In this tilted posture, the dew runs down their backs and settles in a tiny, life-giving blob of water around their mouths. The beetles drink deeply and then start the daily search for food. Among desert-living reptiles, some puff-adders are said to obtain all their water in a similar manner.

Desert birds suffer the same problems, but are able, from time to time, to fly off far away to find water-holes. In the Australian deserts after a rare storm, huge swarms of these birds, almost like locusts, descend on small pools or ponds and compete there for the precious liquid. During the rest

Desert rodents must find special ways of obtaining water. The American pack rat gets all its water from the insides of cacti, having developed a digestive system that can neutralise the poisons present in cactus juice.

The darkling beetle living in the dry desert must take its water from the dew that forms each morning. It does this by emerging at dawn and standing with its rear end higher than its head. This forces a droplet of water to form near its mouth and it can then drink.

Wild budgerigars, living in the arid regions of Australia, can go for long periods without drinking, but when the rains come they flock to pools and drink greedily. In temperatures of nearly 90 degrees Fahrenheit, they can go for thirty-eight days without water, and at lower temperatures they can survive for considerably longer.

Sandgrouse provide water for their chicks in a most unusual way. The male parent flies to water where he soaks his lower feathers until they are sodden. He then returns to his offspring and invites the chicks to run to him. The chicks push underneath their father's body and suck the water from his breast, changing places with one another until all the moisture is gone.

of the time they conserve water as best they can, having unusually dry faeces. Anyone keeping pet birds in aviaries can easily tell whether a particular species comes from a wet, lush environment, or an especially arid one, simply by examining the birds' droppings. Zebra finches and budgerigars, two of the most popular cage birds, are both dry-country species with very dry droppings – and this is part of their appeal, as they have much less odour and make far less mess than, say, typical fruit-eaters. Water is so important to zebra finches that they will only breed after the rains have come. They have no fixed breeding season, but come into full reproductive condition within days of a storm and a good drink. In captivity a water spray and some green grass can do wonders for breeding success with this species.

Breeding in desert conditions is difficult, especially for the growing young that cannot yet fly off to the distant water-holes. Sandgrouse solve this problem in a novel way. When the male parent drinks, he stands in the shallow water with his breast feathers fluffed out so that they become soaked. Saturated with water in this way, the bird then flies off quickly to its nest, where he allows the nestlings to suck the moisture from his feathers. The nestlings have no other way of obtaining water and are so fixated on this method of taking it that in captivity they refuse a water dish and will only suck water from a moist substance such as wet cotton.

Apart from the deserts there are two other environments where drinking water is not readily available. In the polar regions all the water is frozen but this does not present too serious a problem. Most cold-country birds do prefer running water for drinking wherever it is available, but in its absence they are quite prepared to eat snow, which quickly melts in their mouths. Small birds have been observed catching snowflakes during a storm, flying this way and that, snapping them up as though hunting insects.

A much greater difficulty is faced by animals that live in a marine environment, where fresh water is largely or completely unobtainable. The sea otter, for example, is known to drink sea water, but how it manages to survive on this is not yet fully understood. Anatomically it does have very strange kidneys and is probably capable of producing highly concentrated urine, like the kangaroo rat, with the excesses of salt disposed of as waste by this route.

Many seabirds employ a different channel of salt-removal. They, too, drink sea water, but have special salt glands on their heads that get rid of the excess. Most of the salt is transported through the blood stream to the nose where it is transferred to the large nasal glands at the base of the bill. Droplets of highly concentrated salt dribble from these glands and with repeated shaking of the bill the bird rids itself of this unwanted but all too available substance. As a result, a seabird that has been drinking a lot of sea water looks as though it is suffering from a bad cold in the head.

Some seabirds find themselves living in areas where there is, by chance, plentiful fresh water and they then take to drinking this in preference. Young seabirds that have been reared in captivity exclusively on fresh water do not develop such large salt-glands as those that have been reared on sea water, so these specialised organs are subject to anatomical improvement if fully used during the growth of the young birds.

In environments where there is no difficulty in obtaining fresh drinking water, there still remains the question of how to transfer it from the pond, pool or puddle to the back of the throat. The elephant's technique is unique. It squirts its water into its throat with its nose. Young elephant calves do not, however, use their trunks when sucking milk from their

Marine animals such as seabirds can drink sea water. They get rid of the excess salt through special nasal glands at the base of the bill, dribbling the saline solution away as if they have a cold. This giant petrel (above) has a conspicuous 'tubenose', a pair of prolonged tubular nostrils evolved in connection with increased salt loss. Marine iguanas (below) experience the same problem and they too have special glands in the nasal cavity which excrete excess salt. Its removal is aided by vigorous shaking actions that throw sprays of salt high in the air.

The giraffe has a specially modified blood system that enables it to lower its head through 18 feet and raise it again quickly without blacking out. It has to be able to do this because its constant fear of predators requires repeated checking of the neighbouring landscape when it is at the water-hole.

mother's nipples. They apply the mouth like any other young mammal. But when they grow older, they follow the parental pattern of filling the trunk with water, curling it round and inserting it between the lips. In the adult, each trunkful contains $1\frac{1}{2}$ gallons of water, and in a full drinking session the elephants will take on board up to 50 gallons – the record capacity for any species.

Giraffes hold a different record. In their case it is not a question of how much they can drink but how far they can quickly raise and lower their heads without blacking out. Every time they visit a water-hole, they must straddle their legs and then lower the head from a position 11 feet above the heart to 7 feet below it. Why this does not throw their circulatory system into chaos, with resultant fainting and dizziness, has long puzzled giraffe-watchers. The answer, not surprisingly, is that they possess a highly unusual blood system. To start with, their heart beat is astonishingly fast for such a large animal. A cow's heart beats 70 times a minute, an elephant's 25 times, while the skyscraper giraffe clocks up 150 beats a minute. This makes for a highly efficient circulation, better able to withstand sudden shocks to its system. In addition, the network of blood vessels supplying the brain is of a special design, including an unusual 'overflow' emergency system and elastic walls that accommodate any sudden increase in blood. The network is also extensive enough to hold sufficient blood for supplying the brain when the head is rapidly raised, preventing any giddiness. One other adaptation is to the structure of the large neck artery, where valves prevent the blood rushing to or from the brain as the head moves down or up.

These specialisations are particularly important because giraffes are at their most vulnerable when drinking. From their straddled position, it takes them a vital second or two to straighten up for fleeing if a lurking predator leaps out at them. Protecting themselves against such a situation, they drink nervously and quickly, raising the head repeatedly to scan the surrounding landscape for clues. One thirsty giraffe was timed at a water-

Elephants are the champion drinkers of the world, filling up with as much as 50 gallons of water whenever they get the chance. They are also the only animals to use the nose to squirt the water into their mouths.

hole and its drinks lasted 40 seconds, then 19, then 17, then 33, then 7 and finally 20 seconds, after which it walked away. Between each drink its head made the high-speed ascent and descent through 18 feet. This was a typical sequence and underlines why this spectacular animal requires such a highly adapted blood system.

For most mammals the actual transfer of liquid is done by lapping. Close observations of a domestic cat lapping milk reveal that it manages to curl the tip of its tongue slightly to create a spoon shape. With this it flicks small quantities of liquid back into the mouth and swallows after every fourth or fifth movement.

One mammal that possesses a strange drinking action is the wolverine. This northern forest-dweller, a kind of giant weasel, makes treading movements with its forepaws when it is lapping up water. This is a response seen in many young carnivores when they are feeding at the nipple. It is a way of stimulating the mother's milk-flow. But wolverines have extended the use of this action and now employ it in any drinking context, all through their adult lives. By treading and lapping simultaneously, they are able to squeeze water up out of soggy, marshy ground and drink it before it sinks out of reach again. It is such a rigid pattern of behaviour that it occurs even when there is no need for it and captive wolverines frequently knock over their water bowls when they are drinking, as a consequence.

In the world of birds there are two methods of drinking: dipping and pumping. In the dip-and-lift action, the bird fills its bill with water, then tilts the head back to let the water run down its throat, repeating the action a number of times until it has quenched its thirst. In the pumping action, the head stays down and the water is sucked into the throat by a peristaltic

Most mammals lap their water, using their tongues to scoop it up. Cats curl the tongue backwards and then flick it forwards, rolling it up slightly into a spoon shape as they do so. They swallow after every four or five laps.

This woodpigeon (above) demonstrates the pumping drinking action. For some reason pigeons and certain other birds have abandoned the more usual avian drinking technique of 'dip-and-lift' in which the water is swallowed by throwing back the head after filling the mouth. Instead they plunge the beak in and suck until they are satiated. The dip (below) and lift (bottom) is demonstrated here by a green woodpecker.

Having a long ant-eating tongue – up to 16 inches in length – covered with extremely sticky saliva makes drinking an easy matter for this Cape pangolin (right).

movement of the oesophagus, a movement which is often clearly visible to the observer.

It is sometimes boldly stated that pigeons pump and all other birds dip, but this is an over-simplification. In reality, quite a number of different birds make use of the head-down pumping action under certain circumstances.

It has been suggested that pigeons and doves employ the pumping technique because it means that they can drink faster and can therefore fly off to safety more quickly. Like all thirsty animals, they are vulnerable to attacks from predators when drinking, and a rapid intake of water is clearly advisable. However, speed is not the only consideration. Alertness is just as important, and that is served better by the dip-and-lift method. Each time the dipping bird lifts its head it gets another split-second chance to check the landscape around the watering-place for clues of danger.

Both avian drinking styles therefore have their advantages and in terms of safety it is hard to choose between them. Some other explanation has to be sought for the favouring of one over the other. A clue comes from the behaviour of certain small grassfinches. These tiny birds have been seen to use both methods, but they employ them in different contexts. When faced with deep water they dip. If there is only a shallow film of liquid they can be seen to push their bills into it and suck. Attempts to dip into it would be inefficient and they are forced to try and draw it up into their mouths by the pigeon-like pumping action.

It seems possible that the pigeon family has descended from ancestors that were driven more and more to drink from extremely shallow sources, so much so that eventually the pumping action became their normal way of drinking and was used even when deeper water became available. Because its speed compensated for its slightly 'unwary' head-down posture, it has managed to survive as the norm for all modern pigeons and doves – a viable alternative to the usual dip-and-lift.

Cannibalism

It comes as a shock to some people to learn that cannibalism is widespread in the animal world. Our horror at the thought of eating human flesh is so strong that we somehow imagine that this activity is extremely rare in nature and only occurs in abnormal conditions when animals are driven to desperate measures. For many species, however, cannibal acts are a frequent and commonplace feature of social life. It is true that in some an increase in stress leads to an increase in cannibalism, but in others it takes place regularly, stress or no stress. Being eaten by your own kind is just one of the natural hazards of being alive.

Sharks provide the most gruesome example of this type of behaviour. For young sand sharks and mackerel sharks the struggle for survival starts early. Although it is hard to believe, these predatory fish produce cannibal embryos. At the onset of pregnancy, the female's body contains a dozen or so small foetuses, but as these start to grow they begin to prey on one another while they are still in her oviduct. Already well equipped with sharp teeth, the bigger and stronger embryos tear to pieces and devour the younger ones. As other eggs start to develop, these too are gobbled up until, in the end, there is only one large, well fed embryo left, stuffed with its unborn brothers and sisters. With no more siblings on which to dine, the survivor then changes its position in the oviduct in such a way that it triggers its own birth.

The zoologist who reported this foetal cannibalism, or *siblicide* as it has been called, discovered the truth when he was dissecting a dead female shark. To his astonishment he was attacked by a 9-inch embryo and bitten on the hand as he explored the interior of the female's corpse. Later, examining the stomach contents of shark embryos, he uncovered the

In the world of frogs cannibalism is not uncommon. Here tadpole may eat tadpole and even frog eat frog.

Male lions may kill and eat the cubs of other males when they take over a pride. The new males then mate with the females and produce cubs of their own. Towards these new cubs, carrying their own genes, they will be model fathers.

sinister reality of what it is like to be a younger brother or sister in the unborn world of these savage marine predators. He aptly described these sharks as species that send forth their young 'well developed, experienced and with a full stomach'.

Sharks are not alone in having this extreme type of juvenile competition. The Alpine salamander offers even worse odds to its unborn offspring. From about sixty fertilised eggs only one to four young survive the pre-natal period. After they have cannibalised their siblings they go through metamorphosis before they are born.

The desert-living spadefoot toads have a slightly more complicated system. When their eggs are laid they develop into two different kinds of tadpoles – one harmless and algae-eating and the other ferocious and predatory. Their jaws and teeth grow differently, fitting their two roles, and the predators then start to feast on the algae-eaters. The cannibals grow huge and grossly fat, sometimes being as long as 7 inches – giant tadpoles by any standards. What happens next depends on the weather. If it rains a great deal, there is a feast of algae and the algae-eaters thrive, but if there is a drought they grow weak and the cannibals devour them all. Either way the species manages to survive.

The young of certain other kinds of animals turn not on their siblings but on their hard-working parents. Once the life-giving and protective roles of the adults have been completed, they are still good for a meal. The female wall spider often dies before her young are ready to leave their cocoon nest. If this happens, they devour her body when they finally emerge. In this way they are provided with an easily obtained first breakfast before they set off to explore the world. A similar occurrence has been observed in certain sheet-web spiders.

The reverse situation, with parents eating their young, is far more frequent. This is common whenever the adults are subjected to undue stress. Their young suddenly cease to transmit the necessary 'protect me' signals and become conveniently available meat. Countless species of mammals, birds, fish and insects have been seen to act in this way, both in the wild and in captivity. With many species it is enough simply for the parents to be disturbed in their maternal or paternal duties. Zoo-keepers know well enough that even the briefest peep at a new litter in a den will be sufficient to put the parents off in extreme cases. Some mothers become so tame that they allow their offspring to be played with and still treat them as infants to be cherished and cared for, but others are much more easily unsettled. In the wild, disturbance by some animal coming near the nest or den may be enough to trigger infanticide and cannibalism. This has been seen in a number of birds of prey, gulls, crows and storks. Among mammals it is common especially in carnivores and rodents.

In one species we now know the exact mechanism by which adults recognise their young. The tree shrew, a small squirrel-like relative of the monkeys, has a special gland on its chest that secretes a 'protection odour'. Young tree shrews that have the scent from this gland rubbed on to their bodies are magically protected from harm. The scent sends out signals that say, 'I am a baby tree shrew and not a tasty meal.' But if their mother fails to mark them with this fragrance they send out a signal that says, 'I am just a piece of meat,' and she gobbles them up without hesitation.

Careful observations have revealed the circumstances which lead to cannibalism in this species. Tree shrews have long bushy tails that are bristled when they are agitated. If conditions are stressful, their tail-hairs stay erect for longer and it is possible to use this as a simple measure of their emotional mood. If they are relaxed and conditions are stable, their tails bristle for about 5 per cent of the time during an average day. They breed happily enough when they are in this calm state. A nest is built in which one to four young are born. They are fed immediately by the female.

then carefully cleaned, marked with her scent, and left to sleep inside the snug safety of the nest cavity. She visits them from time to time to feed them and is generally an attentive, efficient parent.

The situation changes drastically when the level of stress rises. If the tail-bristling increases fourfold, to 20 per cent of the time, the female may still mate and give birth successfully, but she fails to secrete the vital, protective scent from her chest gland. Unmarked with her fragrance, the young do not smell like young and within a few hours they are eaten.

Close scrutiny showed that this was not a case of an agitated mother refusing to mark her offspring with her oily scent. She had no choice, because when her state of stress rose to this high level it automatically stopped her gland secreting. Even if she rubbed her chest against her babies, it would have no effect.

The overriding importance of the protection-scent is confirmed by a simple test. If young tree shrews that have been properly marked with scent by their own mother are taken from their home nest and placed in the nest of a stressed mother, she does not eat them. In other words, she will eat her own unmarked young but will not eat the marked young of a stranger. It is hard to imagine any scent signal being more emphatic than that in its operation. And surprisingly it proves that the protection-odour of this animal is not a personal scent of each mother but a 'species-scent' that crudely labels a nestling as a 'young tree shrew' rather than as 'my own baby'.

It is difficult for us to understand this type of labelling, where a single property can mean the difference between baby and meat, but signals of this kind are common in the animal world. A similar one is employed by the common house mouse. If a pregnant female smells the scent of a strange male – a male mouse that she has not met before – she suffers a spontaneous abortion. Her whole breeding cycle is halted and she reabsorbs her embryos. Known as the Bruce Effect after its discoverer, this phenomenon has the important function of preventing over-population in colonies of mice. If a colony is so overrun that the resident males cannot keep strange males out, then a dangerously high population level has been reached and the brake that the Bruce Effect puts on stops further increases. It is intriguing to consider what would have happened to human society if women experienced natural abortions under similar conditions.

Stressful and overcrowded environments lead to egg-eating and infant-

The female praying mantis devours the male as he copulates with her. By eating his head first, she removes the inhibitory centre that limits his sexual response. Once he has lost his head, he has no choice but to continue pumping sperm into her body. Some males make the fatal error of approaching the female from the front. This nearly always leads to an attack before they have had a chance to mount the female and start copulating.

Many insects engage in cannibalistic activities. Certain kinds of male flies suffer mating dangers similar to those of the mantis, and occasionally male will also eat male, as can be seen in this lethal encounter between two scorpionflies.

eating in many species. In a pond containing a few goldfish, hundreds of small fry may hatch and many will grow to adulthood. But in the same pond that has become full of goldfish, most fry are eaten and hardly any manage to survive to swell the population further. This mechanism keeps populations down to a level where there is enough food to go around. But space rather than food is usually the key factor. If animals are kept in an artificially limited living area and are provided with an endless supply of food, they still suffer if their numbers rise too high. Crowding alone is enough to cause a breakdown in the breeding cycle and cannibalism of the young. This has been demonstrated vividly in a laboratory experiment with rats, where the whole social structure collapsed in chaos as the numbers reached very high levels. And it has been a regular feature of the appalling conditions under which domestic animals such as pigs and chickens are kept on modern factory farms. Instead of responding sympathetically to the obvious indications of high stress, the reaction of the factory farmers has been to subject their animals to systematic mutilation – such as burning off the beaks of chickens – to prevent the animals attacking one another. Such is the state of farming today.

There is one special circumstance in which young animals may be cannibalised by adults who are *not* suffering undue stress. This occurs when the young ones' father is deposed from his dominant status and is replaced by a new male. The newcomer's first reaction in such a case may be to kill and devour the juveniles. The impact of this onslaught is to bring the females into heat again. They can then be made pregnant by the new male and the next batch of young will carry his genes. This strategy is well documented in wild lions, when a younger and stronger male takes over the pride of an older male. A similar pattern has been observed in a number of other mammalian species, including bears, hamsters and langurs, and in various fish.

In addition to cases where *young eat young*, *young eat adults* and *adults eat young*, there are also a few spectacular instances of *adults eating adults*. The most famous example of this is undoubtedly the lethal mating act of the praying mantis. The female attempts to devour the male as he copulates with her. At the start of the encounter the male approaches with great stealth from behind her. Edging forwards, he can take as long as an hour to come close. Once there he clasps her and starts to mate. He may manage to complete the act and jump to safety, but the chances are that she will bite off his head and eat it while he is still coupled to her. Far from interfering with his mating act, this decapitation actually promotes it, and ensures that the male will continue to pump his sperm into her body. The reason for this is that the inhibitory centre in the male's nervous system is

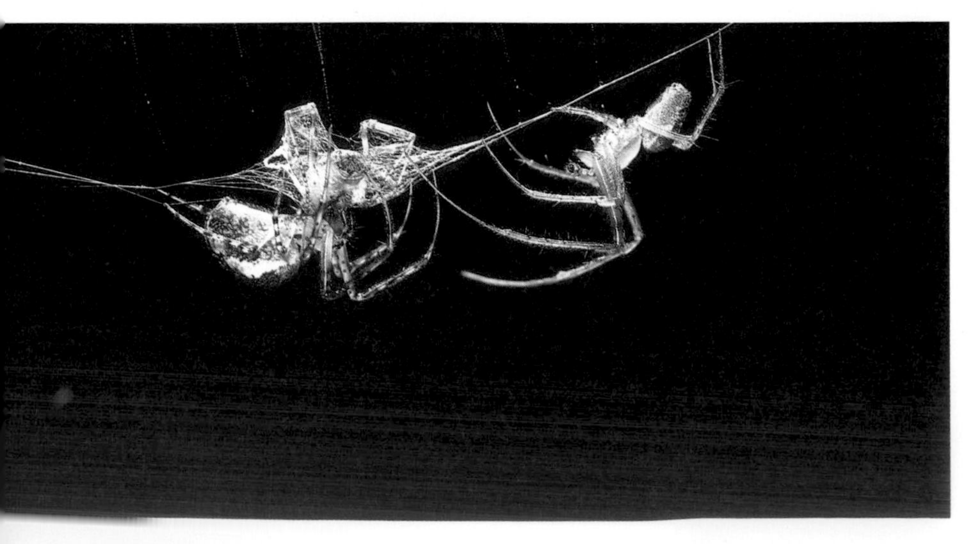

Females spiders will often try to eat the males with which they have just copulated. The males use a variety of devices to avoid this, including tying the female down with silken threads and presenting her with a gift-wrapped meal (left). These strategies do not always work and sometimes (opposite) the male himself becomes the meal.

situated in his head and once he has lost his head he has automatically lost all his sexual inhibitions. His headless body simply cannot stop pumping sperm and the female is assured of fertilised eggs. She continues to eat the male from head to tail as he continues to mate until she reaches his abdomen, by which time he is in any case fully spent. The female is now not only well fertilised but also well fed and can lay a richly nourished batch of eggs.

Other insects enjoy similar nuptial feasts, but the females usually wait until the male has finished copulating before devouring him. This pattern of cannibalism has been observed in certain crickets, grasshoppers, ant-lions and ground beetles. In empid flies there is an additional refinement. Males of certain species wrap up a small food parcel and present it to the female as a wedding gift. The wrapping is important because it takes her a long time to unravel it and get at her prize, and while she is preoccupied the male has long enough to mate without being eaten. In some species the male's gift is less than generous, because he sucks dry the prey before he wraps it up and gives it to his intended mate. For him, when the wrappings come off it is time to leave. In yet other species, the male's gift is further reduced. He carefully wraps up nothing at all and gives his female a large but completely empty cocoon to unwrap. With tactics like these, speedy sex becomes the order of the day.

Male spiders employ even more varied techniques in the struggle to avoid being their own wedding breakfasts. In most cases the females are much larger than they are, and are so intensely predatory that they attack and kill almost anything that moves, without waiting to check on the nature of their prey. Some males overcome this problem by making a series of special visual signals from a respectful distance. Others tap out a code of distinctive vibrations on the female's web, tweaking the threads in a rhythm that lets her know that they are males and not meals. Still others adopt the bearing-of-gifts technique, offering insects that will keep the females busy. Some males are more assertive and put the female into a kind of hypnotic sleep by biting her in a special way, or by tying her down with silken threads and then mating with her when she is well wrapped up. Many of these tactics work and the males live to mate another day, but from time to time they fail and the female satisfies her protein hunger by sucking her mate dry – an undeniably efficient way of giving her newly fertilised eggs a nutritious start in life.

Although cannibalism has been recorded in at least 138 different species of animals – and that excludes cases where stress or overcrowding are operating – it would be wrong to exaggerate its importance in the world of animals. Much more impressive is the remarkable resistance most animals have to making a juicy meal of some highly vulnerable member of their own species. Hungry parents must frequently be tempted to eat their own eggs or young, but their parental urges inhibit them. A most poignant example of this concerns a male jewel fish who was busily rounding up his small fry, taking them into his mouth and carrying them back to the safety of his nest. He had just gulped in yet another of his straying offspring, when he spotted a juicy worm sinking through the water. Unable to resist, he gobbled that up too and now faced a terrible dilemma. In his cavernous mouth he now held one of his precious offspring and a much needed prey. If he swallowed he lost his fry, but if he disgorged his mouth-contents back into the nest he lost his prey. For a moment he hesitated, his fin movements reflecting his intense conflict. Then, his problem solved, he gently spat out both the fry and the worm. As they floated briefly in the water in front of him, he snatched up the worm and quickly swallowed it down. He then gulped up his fry again, swam back to his nest, and spat the young fish back to safety.

Tool-using

WITH A LITTLE encouragement, captive chimpanzees can learn to use tools in a remarkably sophisticated way. They can insert and turn keys to open locked doors, they can place coins in slot machines to obtain food, they can aim balls accurately to dislodge small objects, and they can handle paint-brushes with considerable dexterity to make simple abstract patterns on paper. The fact that they can do all this and also operate all kinds of complicated machinery in laboratory tests is impressive but it tells us little about how tool-using takes place in the wild. To understand that, a great deal of patient animal-watching was necessary in the forests of tropical Africa. Surprisingly, wild chimpanzees were only studied in detail in the past few decades. Previously they were thought to be too difficult to observe at close quarters in their natural environment and were ignored in favour of the more accessible species such as baboons. But then a new breed of field-worker appeared on the scene in the 1960s and we now know of at least four ways in which the manipulative skills of the chimpanzee are put to work without any human aid or influence.

Wild chimpanzees will use broken branches as clubs or as missiles with which to attack predators. When feeding, they will also use stones to crack hard nuts, and this is not a casual activity. The stones and the nuts in question were not found together. Each had to be collected separately and then brought together at a suitable spot. A whole armful of nuts was carried to a flat rock by one ape, who then proceeded methodically to crack them one by one, shell them and eat them. When thirsty they have been seen to fashion a sponge from a wad of leaves and dip it into the small pools of water that form in certain tree-forks. Soaked with water, the sponge is then raised to the lips and the liquid sucked from it.

The most intriguing use of tools by wild chimpanzees is, however, their fashioning of special probe-sticks with which they can obtain live termites or ants from the interior of their nests. Chimpanzees, like so many species that are thought of as exclusively herbivorous, in reality have a strong craving for animal supplements to their diet. One source of animal food is live insects and the apes regularly seek out termite hills or ant nests and break small holes in the hard structures with their strong fingers and sharp nails. But deeper probing is needed. This is achieved by taking small twigs

Wild chimpanzees feed on termites and ants by probing into their nests with long thin sticks. They not only use these tools, but also fashion them by removing side-projections and trimming them to length. In other words they are not only tool-users but also tool-makers, a capacity once thought to be uniquely human.

Carnivores have been observed to use hard surfaces to open eggs, but some are more successful than others. The mongoose has evolved a special flinging action with which it hurls eggs at nearby rocks, smashes them and eats the contents. Some species throw the eggs downwards from a bipedal starting point, but others, such as this banded mongoose in Kenya (above, left), throw it backwards. The lion, although attracted to ostrich eggs, has no special method of opening them and finds them frustrating objects to deal with (above).

and carefully improving their shape by breaking them to a manageable length and removing all side-projections. Once a smooth, unbranched tool has been manufactured, it is laboriously inserted into the hole and pushed right in. This action is done with great concentration and restraint, and then the animal gently withdraws the tool, now bristling with angry insects, and wipes it sideways across its mouth. In this way, the insects are scooped up and quickly swallowed. Again, there is nothing casual about this tool-use. Where no suitable tools are available close to a nest, chimpanzees have been seen to travel for up to half a mile to locate some. Once an efficient probing-stalk has been made, it may be carried around from nest to nest as its owner searches for a good site for its 'termite-fishing'. And if a favourite probe is broken, it is repaired and then reused.

The most important aspect of this nest-probing activity is that it not only involves tool-use but also tool-making. As recently as 1961 it was boldly stated that 'man may be distinguished as the tool-making primate', and our species was referred to as 'man the tool-maker', but, like so many imagined differences between us and the other animals, this separation can no longer be accepted.

Chimpanzees are not the only wild mammals to employ implements of some kind. Certain monkeys have been seen to crack open crabs, using hard stones to batter their armoured bodies, and thick-shelled nuts have been split apart in the same way. The exceptionally intelligent capuchin monkeys from South America have been observed hammering nuts against hard surfaces and crushing them between stones.

Primates are not alone in this. Several carnivores enlist the use of hard objects to help them break in to protected food objects. Mongooses and skunks have special body movements that enable them to fling birds' eggs against rocks, as a way of cracking them open. Different species employ slightly different techniques. The long-legged marsh mongoose rears up in the air, holding the egg in its front paws, and then flings it violently downwards. The grey mongoose hurls the egg backwards, between its hind legs, at a vertical rock-face. The spotted skunk also hurls the egg backwards, but helps it on its way with a kick from one hind leg as it does so.

The sea otter consumes large quantities of shellfish and uses a pebble as a tool to crack them open. It employs two different techniques. Shellfish

The sea otter dives to collect a flat stone and a bivalve mollusc such as a clam. It then surfaces, flops over on to its back, places the stone on its chest, takes the clam firmly in both front feet, raises its arms above its head and starts to batter the shell against the surface of the stone. After a number of blows in quick succession, the shell is cracked and the otter can enjoy its meal.

clinging to rocks are battered with the stone in a straightforward way, but shellfish brought up from the bottom are treated in a more unusual manner. The otter dives down and collects a flat stone and a bivalve such as a mussel. Coming to the surface again, it rolls on to its back and places the flat stone on its chest. Holding the mussel in both front feet, it proceeds to hammer it repeatedly down on to its portable anvil until the shell breaks open and the contents can be devoured.

This shell-hammering by sea otters is extremely hard work and only makes sense because it avoids the necessity of the animals carrying their shells ashore to open them. One otter observed over a 90-minute period managed to break open a total of fifty-four mussels, using no fewer than 2,237 blows on its flat stone. Not surprisingly, daily feeding of this kind causes considerable wear and tear to the chest beneath the anvil, and is believed to be the cause of a variety of chest complaints in this species.

Because tool-use only occurs in certain colonies of sea otters and appears to be completely absent in others, it was originally believed that this must be a learnt pattern of behaviour, spreading through one group but failing to be invented and exploited in another. This is no longer thought to be the case. Closer examination has revealed that in those colonies where it is absent the local shellfish, or other marine food, is weak enough to be crushed with the otters' teeth and jaws. Tools are not needed and the use of an anvil is simply not called upon. The important clue came from an intermediate colony, where the typical food was weak enough for adults to be able to crush it in their jaws but too hard for the younger otters. In that colony, the young otters alone employed the stone anvil technique, demonstrating that this is not, as previously imagined, a feeding pattern learned by imitation of parental activities, but a behaviour which occurs automatically as an inborn response to the presence of a difficult prey object.

This does not mean to say that individual learning does not assist in the improvement and development of the basic hammering reaction. Some adults, for example, in colonies where all the otters are active tool-users, have become far-sighted, retaining their favourite anvils for future use. A particular flat stone is kept for repeated hammering sessions, being brought back each time to the surface with the newly acquired mussel. The only mystery with this practice is how the sea otters hang on to the stone while diving. Close observation has revealed the answer: they dive to the bottom each with their anvil tucked neatly into one armpit.

Outside the primates and the carnivores, there is little mammalian tool-use. Elephants and horses have occasionally been seen to pick up a stick, hold it in the trunk or mouth, and use it for back-scratching. In this way they can reach a particularly irritating spot without having to lie down and roll on the ground or get into an awkward position and rub against a tree or a rock. But despite the generally impressive intelligence of mammals there are virtually no other instances of them using natural objects as implements. To find additional examples, it is necessary to turn to the birds.

Perhaps the most famous of all tool-using animals, and the first to be studied carefully, is the woodpecker finch from the Galapagos Islands. This small bird lacks the long, probing tongue of the true woodpecker, but makes up for this deficiency by using a sharp, slender twig or a large spine from a cactus plant. In its search for insects hidden in the cracks and crevices of branches, the finch holds the tool lengthwise in its mouth, like an extension of its beak. Finding a likely spot, it pushes and prods the twig or spine into the cavity until it either impales a concealed grub or disturbs it enough to drive it out into the open. It then drops the tool and quickly picks up and eats its prey.

The woodpecker finch of the Galapagos Islands employs a sharp twig or a large cactus spine as a tool when searching for grubs that live hidden inside wood or for insects that conceal themselves in cracks and crevices in branches. It impales them or uses the stick to lever them out. This finch was one of the first known animal tool-users.

Where wood-boring grubs are present, the woodpecker finch varies its technique. It begins by pecking a hole in the branch until it reaches an insect burrow. It then flies off and selects a suitable implement and returns to the freshly excavated aperture. Probing the tool deep down into the hole, it then either drives the prey out or, failing that, uses the tool as a lever to prise it up to the surface.

The most surprising observation with this small bird, however, has been that, like man and the chimpanzee, it is not only a tool-user but also a tool-maker. If there is no suitable probe-stick or spine readily available, it will select a larger twig and then set about cutting it down to size, removing small side-projections in the process. This is precisely what the chimpanzees do when improving twigs for their termite-nest probings and, in this respect at least, the ape brain and the bird brain are on a par.

Another extraordinary example of avian tool-making is found in certain of the bowerbirds. The satin bowerbird and several other species select small pieces of wood and shape them into 'brushes' with which they paint the walls of their elaborately constructed bowers. They take up a mouthful of charcoal or fruit-juice, mix it with their saliva and then dribble it down the 'brush' which is held between their slightly open mandibles. In these cases, tool-use is connected not with food-finding but with the sexual displays of these remarkable birds.

Many birds use hard surfaces as immobile implements in their attempts to obtain food objects that are protected by a strong outer casing. The song thrush at its anvil, hammering a snail-shell against a rock or a stone, is a familiar sight in European gardens. The snail is held in the bird's bill and flicked downwards with a rapid, swinging motion, so that the apex of the shell is cracked open. It takes several swipes with considerable force to achieve this, and a few more to enlarge the crack. This damage snaps the columella, or central axis, of the shell and renders powerless the columella muscle with which the snail retracts its body inside the shell. There is then nothing left to prevent the mollusc's eviction and a few more flicks achieve this, the naked snail being thrown out of the shell through the hole that the thrush has made. The bird then picks up the snail in its bill and proceeds to wipe it repeatedly on the ground, removing as much mucus as possible before swallowing it.

The song thrush employs an anvil against which it can smash open snails. It holds a snail by the lip of the shell and cracks apart its apex. This destroys the snail's hold on its home and it is easily evicted and its soft body devoured.

The song thrush is not the only bird to use tools in this way. Anvils are employed in feeding on hard-shelled prey by other species, such as the white-winged chough, the buff-breasted pitta and the stagemaker bowerbird. Gulls and crows have also been observed to use rocks as shell-

The Egyptian vulture has a technique of picking up a stone, the largest it can hold in its beak, and throwing it down as hard as it can on to an egg. Using this action it has even been able to crack open an ostrich egg, although the tough shell often takes several direct hits to shatter it. With smaller eggs, the vulture picks up in its beak the food object itself and flings it to the ground.

(Opposite) The archer fish employs droplets of water as missiles, using them to shoot down insects resting on overhanging vegetation. It is accurate up to 5 feet away and can sometimes manage as far as 10 feet with its 'spitting' actions. The most remarkable aspect of this behaviour is that the fish is capable of making corrections for the surface refraction of the water.

smashers, but in their cases they let gravity do the work for them. They fly up high above a rocky surface and then drop the food-objects – shellfish, eggs and sometimes tortoises – down on to the 'anvil' below. Bearded vultures have also been seen to use this method of food-preparation – often dropping bones and, again, tortoises – but there is another species of vulture, the Egyptian vulture, which has a more impressive technique.

Ostrich eggs fascinate and frustrate the Egyptian vulture because, although they look exactly like other eggs and therefore hold the promise of a good meal, their shells are too thick and too strong to succumb to the normal assault of simple pecking with the beak. This difficulty is overcome in a spectacular way. The vulture seeks out a suitable stone – large enough and heavy enough to crack the ostrich egg, but not too big to hold in its beak. It picks it up and then, standing in front of the egg, pulls itself up to its maximum height with its neck fully stretched and flings the stone down as hard as it can. It strikes the egg in about half of its attempts and it takes between six and twelve direct hits to crack the thick shell open. If ostriches have laid their eggs in a sandy, stone-free zone, these vultures will travel far and wide in search of suitable missiles. They have been known to carry a stone a distance of 3 miles in extreme cases, revealing that this is not a casual, spur-of-the-moment response but one that is capable of involving prolonged concentration.

Ornithologists have been able to uncover cases of tool-using in at least thirty species of birds. The following are a few more of these examples. Carrion crows have been seen to hold an acorn under their feet, reach for a small stone and then use it to hammer the nut open. Shrikes employ sharp thorns on which to impale their prey. Many small birds will pick up an ant in the bill and wipe it through their wing-feathers, squeezing the formic acid out of its body as they do so. This is thought to be a form of pest-control, with the acid killing off the birds' troublesome ectoparasites. In such instances the ants can be thought of as 'living tools'. Green herons employ their own feathers as bait when catching fish. The birds drop a feather on to the water near small fish. The unsuspecting prey respond to this novel object by investigating it as possible food. When they nibble at it the heron stabs them and eats them.

Isolated observations suggest that there is probably a great deal more tool-use going on in the world of birds than has previously been imagined. The little known Galapagos mangrove finch has recently been observed to behave in the same manner as its well known relative the woodpecker finch, and several other species have also been recorded as probable tool-users, although more accurate information is needed. And in captivity, the list grows even longer, with many (perhaps bored?) species of parrots, crows and birds of prey, employing small objects to groom themselves, obtain food, or perform tricks.

Among fish the only convincing tool-use example concerns the species that employs water-droplets as weapons. The archer fish that inhabits the mangrove swamps of South-east Asia spends a great deal of its time spitting its missiles at unwary insects resting on the overhanging vegetation. It is capable of shooting the small liquid bullets with great accuracy over considerable distances. It has almost pin-point precision up to 5 feet but in exceptional cases has been able to reach prey 10 feet away with its bombardment of water-drops. Its technique is to use its mouth like the barrel of a rifle. There is a groove running along the top of the mouth and when the fish's tongue is pressed up against this it creates a narrow tube through which the water can be expelled under pressure. The pressure is produced by a sudden contraction of the animal's gill-covers. The most amazing feature of the archer fish's performance is that it is capable of allowing for the surface refraction of the water, because although its lips are just out of the water when it fires its droplet missiles, its eyes are firmly

Shrikes employ thorns as implements when feeding. A carcass is impaled on a thorn as a method of temporary storage and some species of shrikes include quite large prey in their list of impaled victims. Frogs, lizards and even mice can be seen skewered on the thorny branches of a busy shrike's tree. Large insects are also common prey, as in this case, where a rufous-backed shrike has impaled a locust. In some areas shrikes have switched to a more modern material, using barbed wire in place of thorny branches.

The weaver ants roll up leaves and stick the ends together with silk, fashioning a container for their nest. The adult ants cannot make silk, however, and must use larvae for this. Some of the ants hold the edges of the leaf firmly together while other ants pick up larvae in their mandibles and squeeze them to force out a little silk. They then start to move the larvae along the edges of the leaf, dragging liquid silk out of them as they go. In other words, they use their own offspring as implements, much as we would squeeze a tube of glue.

below the surface. This means that it must somehow correct for the distortion created by looking from one medium into the other. It seems that this is something that has to be learned by each individual fish. When they are young, archers spit at almost anything and with little directional sense. Then, as they grow older, their aiming ability gradually improves until, as adults, they have become experienced and expert marksmen. In other words, the spitting is inborn, but the aiming is learned. To demonstrate their fire-power, a small shoal of archer fish kept in a public aquarium was fed in a novel manner. Each day a mass of finely ground raw hamburger was thrown against the glass wall of their tank, above the waterline. It stuck there and the little fish immediately began eyeing the tempting meat clinging to the vertical surface, just out of reach. A few of them tried to jump up and snatch some of it, leaping as much as 12 inches clear of the water. But then the whole shoal moved in and started their droplet-missile barrage. Some used single shots, with pauses in between; others employed machine-gun bursts. In fifteen minutes the shoal had washed $\frac{1}{3}$ pound of raw meat off the glass into the water below. As tropical-insect controllers they are clearly in a class of their own.

From the world of invertebrates there are two notable examples of tool-use. The first concerns a female digger wasp that builds a special nest in sandy ground. This consists of a vertical shaft into which she will lay her egg. But before doing this she provides her offspring with a future meal by paralysing a caterpillar and stuffing it down into the burrow. As a finishing touch she has to plug the nest, or the caterpillar's body might attract other hungry mouths. She does this by filling the hole with ordinary grains of sand, but there is a danger that these will become disturbed if they are too loose. The solution is to tamp them down firmly and to do this the wasp employs a tool. She searches for a small pebble or stone, bigger than the ordinary sand grains but not quite as large as her own head. Holding it in her mandibles, she uses the pebble to hammer down the sand blocking the nest entrance. She continues with the pounding until the entrance has been made sufficiently firm, and her maternal task is completed.

The second example is perhaps the most extraordinary of all the cases of animal tool-using. Weaver ants had for some time been known to make nests by rolling up leaves and fixing the edges together with silk. But as this was accomplished by adult ants that produced no silk themselves, how they managed it was something of a puzzle. The answer to the mystery is that while a row of adults holds together the leaf-edges other ants pick up ant-larvae in their mandibles and use them as living tools. By squeezing them slightly they force them to exude liquid silk, and once a drop of this has been placed on a leaf-edge the larva is pulled along, extruding more and more silk and fastening the ends of the curled leaf together. The adult ants employ their larvae in this way rather as we would squeeze a tube of glue. Other larvae are used like weaving shuttles to line the interior of the nest, being pulled back and forth until the task is completed. It has been claimed that this is enforced child labour rather than tool-use, but the larvae are so passively involved that referring to them as blunt instruments seems somehow more appropriate.

On the question of the definition of tool-use, it has sometimes been argued that a few of the examples mentioned here are marginal. Certain authorities consider it essential that a tool should be a manipulated object that is used as an extension of an animals' body. This would rule out cases such as the dropping of a prey on to a hard rock surface, but would include the dropping of a hard stone on to a prey. This distinction has not been made here because, in human terms, the blacksmith's immobile anvil and his mobile hammer are both important implements used in fulfilling his task. So it is for animals, and the narrower definition has little to recommend it.

Conflict Behaviour

When an animal is in a state of internal conflict, its behaviour becomes intense and complicated. Torn in two different directions at the same time, the performer finds it impossible to obey either of the two opposite pressures. Each cancels the other out, but not in a way that leaves the animal calm or resolved. On the contrary, at such times there is tension and agitation, both in postures and movements. The strange actions performed provide a basis for the evolution of animal body language.

Because moments of inner conflict are times when moods are clashing and fluctuating, they are also times when it becomes urgent for animals to communicate their emotional states to partners, mates, rivals, parents, young or companions. It follows that, during the course of evolution, many responses to conflict have become modified or exaggerated as special rituals. This process of ritualisation sees the rigidification of the actions that erupt when opposing forces prevent the simple, smooth running of single-minded behaviour sequences. And this rigidifying of behaviour follows certain set patterns and rules that help us to understand the true nature of greeting displays, threat displays, courtship ceremonies, parental activities and infantile behaviour.

Seven key patterns are: ambivalent actions; intention movements; redirected activities; displacement activities; re-motivating actions; cut-off; and autonomic signals. By briefly identifying each of these themes it will become much easier to unravel the complexities of even the most bizarre and apparently mysterious animal dances, displays and disputes.

1 Ambivalent Actions

When an animal is strongly stimulated to retreat and at the same time is equally strongly stimulated to advance it should, in theory, remain exactly where it is. The perfectly balanced urges to go forward and go back should leave it rooted to the spot, unable to move a muscle. In practice, of course, such precise balancing is highly improbable except for a split second. What happens in reality is that one urge starts to get the upper hand and the animal begins to edge in that direction. If the conflict is, say, between fear and aggression, and fear increases slightly, the animal will start to pull its body backwards slightly. As it does this, the small retreat makes it feel safer because it is now farther away from the source of its fear. This in turn makes it feel more assertive and it edges forward. Coming closer to its rival, it now feels more frightened again and once more has to retreat a little. In this way a pendulum effect develops, the animal edging back and forth in an agitated manner.

This ambivalent action can easily become the basis for ritualised dancing displays. During the course of evolution, the back-and-forth jerking becomes smoother and more rhythmic. The precise form of the movements will vary from case to case. For some species it may mean a rhythmic pivoting from side to side, while for others it ends up as a vertical bobbing, a zigzag darting, or some kind of strange twisting and turning. There are as many possibilities for ambivalent animal displays as there are for human dance-steps on the dance-floor. Some species waltz, some rock-and-roll and others do the twist. And frequently they add their own species-music to the dance.

Like all conflict responses, these ambivalent actions may occur whenever two contradictory stimuli reach the animal. In the example given, these were fear and aggression, but they could just as well have been fear and sexual attraction, aggression and sexual attraction, fear and parental care, aggression and parental care, and so on. Sometimes there may be three urges operating simultaneously. In one species, different contexts may create different types of display; in others, the same display may suffice for more than one type of encounter.

When animals are about to depart they make small intention movements indicating that they are preparing to initiate locomotion of some kind. This little owl crouching for take-off is performing an intention movement that is easy to understand, but the crouch can evolve into elaborate head-bobbing or, as in the case of this garganey duck, into the tossing-back of the head.

2 Intention Movements

These are closely related to the last category, but only one of the two urges is expressed. If a bird is frightened and has a strong urge to flee, but also wishes to stay where it is for some reason, it may start to open its wings for flight, or fan its tail for balance, or dip its tail for take-off, or bend its legs for push-off, or crouch its body to spring up. These and other such 'get-ready' movements reveal its intention of flying away, but something prevents the follow-through into actual flight. Only the very first stage of the whole movement is actually performed, before the bird checks itself and halts its departure.

These curtailed movements act as vital clues to other birds, telling them of the performer's intentions. According to the form and direction of the actions, so it is possible to detect what kind of outcome would have developed had the bird allowed itself full expression. Some forward-facing ones suggest an imminent attack, while sideways ones look more like a lateral retreat. Again, during the course of evolution these intention movements have often become exaggerated and made more conspicuous. A small, erratic bobbing of the head or flicking of the tail becomes amplified into a more vigorous, more regular bobbing or flicking. These ritualised intention movements become signals for all members of a particular species and probably differ from those of closely related species. In this way conflict responses can aid species isolation. Each type of bird will have its own 'private' body language based on the many different ways in which simple 'starting actions' can be exaggerated and distorted. With one bird species, the tail flick becomes the all-important clue, with another it is the head-bob, and with yet another it is the leg-crouch. These actions will now be not only different from one another and different from the simple 'ancestral' form but will also eventually become more and more commonly displayed and increasingly conspicuous. In the end they virtually substitute for a real 'take-off' movement, the full take-off for flight being reserved for extreme conditions.

Intention movements of flying towards an enemy, of the kind seen here in a greenfinch (above) can become developed into startling frontal threat postures, as displayed here by a great horned owl (opposite)

3 Redirected activities

When two animals meet and feel antagonistic towards one another, they frequently find that their mutual fear is too strong to permit them to carry through a complete attack. They stand angrily face-to-face, their state of inner conflict making them impotent to act. Under such conditions they may find a way of expressing their pent-up urges by turning to one side and attacking a harmless third party. This redirection of aggression is a major problem in human society. A man arriving home angry and frustrated after a bad day at the office may vent his feelings on his wife and children, who are in a weaker position. Failing that he may kick the cat. And if he loves his family and his pet too much to take it out on them, he may start smashing some inanimate object. This same principle works for other species. Indeed in some gulls it is so common that it has led to a special grass-pulling display. When two of the gulls meet, are stimulated to attack one another but cannot bring themselves to do so, one of them may suddenly turn to the side and start tearing aggressively at some long grasses, as though it were ripping at its opponent's feathers. Its anger is redirected from the body of its rival to the harmless grass-stems.

4 Displacement Activities

During the most intense moments of conflict, a strange phenomenon often occurs. Animal-watchers have for years been puzzled by the fact that right in the middle of a dramatic encounter an animal may suddenly pause and perform what appears to be a totally irrelevant action. An intensely courting bird may rapidly wipe its beak; a threatening fish may quickly dig

When animals are in an intense state of conflict they sometimes pause and perform small, seemingly irrelevant actions, such as grooming themselves, as this kangaroo is doing (below), or preening their feathers, as this mandarin drake is doing (right). Displacement yawning also occurs when there is tension, even in fish (bottom).

in the sand; a cat worried by a strange prey animal may break off its attack and start grooming its fur. These brief, cursory actions, often incompletely performed, have been called displacement activities because it is as if the behaviour output of the animal has been momentarily displaced into another channel of expression.

Whereas in the case of redirected activities, the animal does not change its action, but applies it in a different direction, with displacement activities it is not merely the direction that is changed but the whole pattern. But then, having allowed itself a moment of release by the performance of some completely different type of activity, it is as if the animal cannot spare too much time or concentration for it, and carries it out in only a token fashion. The way it has been modified can be measured in some cases. Small finches that wipe their beaks in the middle of intense bouts of courtship do so in a highly abbreviated way. In true beak-cleaning, when the bird is dirty after feeding, it always scrapes the beak on the branch several times with a characteristic speed. In displacement beak-wiping it often makes the movement in mid-air, without even touching the branch. It also employs fewer movements and it performs the actions more quickly.

This same species of finch – the Australian zebra finch – may also break off its bouts of courtship or fighting to perform displacement preening, stretching, feather-shaking, head-scratching, yawning, sleeping, feeding, nesting, and various other activities. Which one is selected in any particular instance seems to be almost arbitrary, although some actions are more commonly interspersed in fighting, while others crop up more often in the middle of courtship bouts.

In some species of ducks the situation is different. There, certain specific kinds of displacement activities have become so closely linked to particular courtship situations that they have become incorporated into specially evolved displays. The drakes employ highly modified preening actions that bring into prominence brightly coloured patches of feathers. The preening has become no more than a single swipe at the feathers, with the bill being aimed at the display patch. In the mandarin, the displacement preening movement has become the touching of a large display feather; in the garganey, the bill is moved to the outside of the bright blue wing-coverts, drawing special attention to them. In the mallard, the evolution of the displacement preening action has taken a different direction. There it has become a sound signal, the drake dragging the tip of his bill across the base of the large pinion quills. This makes a loud *rrrr* sound, easily audible to the courted female.

The value of these displacement activities to the companions of the animal that perform them is that they reveal a state of finely balanced conflict. This makes them useful indicators of mood and it is not surprising that they have become important signals in animal communication.

5 Re-motivating Actions

If a weak animal wishes to approach a strong one it may achieve its goal by changing the mood of its dominant companion. It can re-motivate it by behaving in a sexual or an infantile manner. The arousal of the companion's sexual or parental feelings helps to suppress its other more aggressive or competitive moods.

During courtship, when there are intense inner conflicts, a bird may suddenly adopt a juvenile food-begging posture. This quickly arouses parental feelings in its companion and these reduce its hostility. This strategy may be so successful that the partner is actually provoked into feeding the begging bird as though it really were a nestling. This has led in many species to ritualised courtship-feeding as an essential part of pair-bonding.

When approaching a companion, an animal may attempt to change its mood and suppress any aggression by performing a non-hostile act. This may be infantile, sexual, or parental, or may perhaps be a grooming invitation. Courtship feeding in birds such as these hoopoes is a common example of such an act. Feeding young chicks is a comforting parental action, but when borrowed by a courting male and performed towards a female it can have a calming effect on her and assist the development of the pair-bond.

At other times, a weak animal may approach a dominant animal and perform the female sexual invitation display. Frightened male birds in a state of conflict may do this and are sometimes mounted by their females as a result. It is also a common device among many monkeys, the weak animal presenting its rump to the strong one, who usually mounts and makes a few token pelvic thrusts. The genders are irrelevant here, and it is possible to see all four possible combinations: male mounts female, male mounts male, female mounts female, female mounts male. All that matters is that conflict tension has been resolved by the performance of actions that are related to a non-aggressive act.

The mood of animals performing re-motivating actions is normally submissive. Their conflict is between fear and the need to stay close to a frightening animal. The exception to this rule is the use of re-motivating acts by dominant animals that wish, for some reason, to stay near to subordinates. This happens occasionally when adult monkeys want to approach frightened juveniles. On such occasions it is possible to witness the strange scene of a huge adult acting submissively towards a tiny juvenile as a way of allaying its fears and preventing it from fleeing.

The most common and widespread of all re-motivating actions is social grooming. When two animals wish to be in close contact with one another but at the same time are slightly nervous of this physical intimacy, they may reduce the tension between them by grooming or preening one another's bodies. This friendly act of mutual aid arouses non-hostile feelings that rapidly suppress any traces of fear or aggression that may be present. As an important development of this process, grooming invitation signals have evolved in a number of species. These small actions can be performed as one animal approaches another. Their primary message is 'I want to groom you', but their more important, secondary message is 'I mean you no harm'. In this way tokens of friendship can grow out of the re-motivating action of social grooming.

In monkeys the most widespread grooming invitation signal is lip-smacking. The mouth is opened and closed rapidly as the animal approaches its companion. This is now a highly ritualised tension-reducer, but it originated in the 'tasting' actions of grooming monkeys, employed when they are removing salty scales of dead skin from one another's fur.

In horses the young foals use a strange snapping action of the mouth when nervously approaching adults. This was originally misinterpreted by

Social grooming is often employed as a remotivating device to defuse a tense situation and enable members of a social group to remain calm and peaceful when close to one another. It is particularly common in primates, such as vervet monkeys (left) and baboons (above).

human observers as an aggressive act because it looked like biting, but in reality it is derived from the nibbling action of equine mutual grooming. Fortunately for foals, adult horses never make this mistake, recognising the action for what it is, a friendly appeasement signal.

Anyone who has kept a pet cat or dog will know the way they invite grooming as a friendly act, and few owners can resist this invitation. The offering of the back of the neck for stroking and caressing often leads to prolonged interactions which owners find as strangely satisfying as do their pets. Many birds invite preening from their mates in a similar way, ruffling the neck feathers and twisting the neck to display the back of the head.

When young foals nervously approach adult horses, they perform a snapping action which, although it looks superficially like a biting movement, is in reality a 'nibbling' display, inviting social grooming.

6 Cut-off

Another phenomenon observed during moments of intense conflict is 'cut-off', in which an animal attempts to survive the stress of an encounter by switching off the incoming messages. If it is determined to stay where it is but is also deeply fearful of the approach of another animal, it may attempt to damp down the level of fear by hiding its face, closing its eyes, turning away its head, or performing any action that will lower sensory input. If it fails to do this, the teetering balance of its conflicting moods may collapse in favour of fleeing and it will be forced to relinquish its position.

It has been argued that the selection of certain types of intention movement and displacement activity for incorporation into special rituals has often been dependent on their value in terms of cut-off. If the intention movement of turning away from the companion serves to hide the face, then it will have more cut-off value than if the animal simply bobs its head up and down while still looking at the companion. In this way it is easier to understand displays such as head-flagging in black-headed gulls, where the two birds simultaneously twist their heads away from one another. These are intention movements of fleeing that succeed as signals because they cut off the source of the fear – the sight of the other bird at close quarters – and permit both partners to relax a little. This relaxation is clearly visible in the demeanour of the birds immediately after they have performed the head-flagging. They are then more able to continue in close proximity to one another.

The displacement preening of ducks has a similar quality. The way the head is turned to swipe the bill through the wing feathers effectively hides the face of the male and again acts as a cut-off action. Many other displays have a similar face-hiding property that helps to reduce the inevitable tension of close encounters during the breeding season.

These two male Barbary macaques are making use of a newborn infant monkey to help to reduce the aggression between them. By paying attention to the baby the males can sit more closely and peacefully together than if they had nothing to distract them. The baby re-motivates them by arousing their parental feelings, and these in turn suppress their antagonistic feelings.

7 Autonomic Signals

Whenever animals experience a strong conflict of moods their autonomic nervous system undergoes dramatic changes. These changes have visible effects that in turn can become part of their signalling systems. To appreciate this it is important to understand the kinds of autonomic changes involved.

The autonomic system is concerned with adjusting the internal workings of the animal's body to fit the requirements of the external environment. When the animal is intensely active physically, as in fleeing or fighting, the autonomic system gears the body up for this action. When the animal is resting, it gears it down again. There are two counterbalancing systems to deal with these two conditions: the *sympathetic* system prepares for high activity and the *parasympathetic* for low activity.

The sympathetic system pours adrenalin into the body and this has a number of dramatic effects:

When an animal is in a state of conflict it may attempt to reduce its stress by simply cutting off the stimulation. It can achieve this by looking away, as these bee-eaters are doing.

- *Alimentary* The storing and digestion of food ceases. Salivation is restrained. Movements of the stomach, the secretion of gastric juices and peristaltic movements of the intestine are all inhibited. The rectum and bladder do not empty as easily as usual.
- *Circulatory* The heart beats faster and blood is transferred from the skin and the viscera to the muscles and the brain, where it is urgently needed. This gives rise to an increase in blood pressure.
- *Respiratory* There are alterations in the rate of respiration. Breathing is both quicker and deeper.
- *Thermoregulatory* The temperature regulation mechanisms are violently activated and sweating increases. Also, hair or feathers are sleeked or fully erected. When sleeked they help to streamline the body for action and they also reduce the thickness of the coat in a way that slightly decreases its insulating properties. When fully erected, the hairs or feathers separate to expose the underlying skin and produce a more powerful cooling effect, but this inevitably also works against streamlining, so that it is an alternative that is not always used.

The parasympathetic system works in opposition to all these changes. It calms the body down, counteracting the extravagant effects of the sympathetic system. Its function is to preserve and restore bodily reserves. Under its stimulation, the digestive processes increase. There is more salivation. Urination and defecation occur more easily. The heart beat slows down and blood is shifted to the skin and alimentary system where it aids in digestion. Breathing slows down. Sweating decreases and the coat resumes its relaxed, thickened condition giving maximum insulation. The hair or plumage assumes a more rounded, fluffed condition.

In the moderately active animal these two systems counterbalance one another. But when the animal is resting or sleeping, the parasympathetic system dominates, and when athletic exercise is called upon, the sympathetic takes over. The most interesting condition, however, occurs when the animal is intensely aroused and is then prevented from expressing itself in vigorous muscular activity. In this strongly thwarted state, which occurs whenever an intense inner conflict arises, the animal is provoked into massive sympathetic responses because of the demands for action. Simultaneously, however, it is equally powerfully provoked into parasympathetic responses because of its inability to carry out the actions demanded of it.

If, for example, a bird is defending its nest against a frightening enemy, it has a strong urge to flee that makes it secrete adrenalin. But its equally strong urge to stay put and protect its nest-site forces it to remain where it is, even though its body has geared it up for the tremendous physical activity necessary for dashing away to safety. This throws its autonomic nervous system into chaos. Contradictory messages keep flying around the body with the result that first one extreme response is given and then its opposing counterpart. So the agitation we see in an animal suffering this type of conflict is not merely muscular – it affects almost all the body systems.

Changes in respiration during stressful encounters may lead to inflation responses in which brightly coloured pouches are distended as special display organs. The most spectacular example of this is the huge red pouch of the courting frigatebird (left). The yellow pouch of a displaying Kirtland's tree snake is also impressive (above).

Stressful encounters may lead to sudden shifts in body temperature. These cause changes in the insulating systems of animals – hairs in mammals and feathers in birds. Much feather ruffling is seen in birds at times of conflict and this type of autonomic display has led to many spectacular forms of feather erection. Seen here are the crest erection of the great black cockatoo (opposite) and the bald eagle (left).

These are some of the characteristic body reactions: dryness in the mouth or excessive salivation; sudden urination or defecation; extreme pallor or intense flushing of the skin; deep and fast or slow and shallow breathing, or irregular, erratically interrupted breathing, leading to gasping or sighing, hissing or rasping; panting, weeping, fainting and sweating; and the sudden raising and lowering of hair or feathers.

It is easy to see how these disturbances, in coinciding with intense conflict situations, have given rise to the following display signals:

Urination and defecation have become adapted as territorial scent-marking devices. Leg-cocking in dogs is the most obvious example.

Skin-flushing has become conspicuous on special bare patches, especially on the face. This is true of various birds and monkeys as well as man.

Erratic breathing had led to hissing displays that are common in reptiles and are also found in some birds and mammals; also to croaking in frogs; and eventually to all forms of vocalisation in birds and mammals, including human speech.

During moments of tension an animal may feel the need to urinate or defecate. This has led to the use of urine and faeces as scent-marking devices. Many territorial mammals display their ownership in this way. Some, such as the rhinoceros (below, left) and the cheetah (below), have the ability to squirt a jet of urine backwards horizontally against a vertical landmark. These marking posts are visited regularly and the scent there replenished.

Respiratory tension has led to a variety of *inflation displays* in which bladders and pouches are dramatically filled with air. Male frigatebirds have huge scarlet throat-pouches that they enlarge during courtship. Other distensible bladders are seen in the displays of birds such as the sooty grouse, the sage hen and the Australian bustard.

Increased sweating during conflict has led to the modification of some sweat glands as specialised scent glands. This is widespread in mammals and has gone hand in hand with increased sensitivity to body fragrances, providing a subtle 'olfactory language' during social encounters.

Hair erection in mammals has become specialised by being restricted to certain parts of the body. In this way displaying mammals can erect distinctive and conspicuous crests, manes, tufts and rump-patches. Feather erection performs a similar function in birds, with the evolution of dramatically exaggerated crests, ruffs, ear-tufts, throat-plumes, chin-tufts, flank-plumes and rump-patches. Frequently the erected hairs or feathers have become longer and more brightly coloured than the surrounding coat or plumage, further enhancing the displays.

Clearly the disturbances in the autonomic nervous system at times of intense thwarting and conflict provide the basis for a vast array of different animal signals. These and the muscular 'agitations' mentioned earlier – the ambivalent actions, intention movements and the rest – offer the raw materials on which the whole of animal body language and communication can be built.

Balancing movements, of the kind shown here by a dipper coming in to land on a rock, are often the basis for bird displays. The spread wings of the physical balancing posture become incorporated into the courtship or aggressive displays.

Typical Intensity

When one animal displays to another it is important that its signals should be clear and sharply defined. If its actions, when communicating its mood, are too vague or too variable, they can easily become ambiguous and cause confusion. As a result, the body language of displaying animals is full of vivid, distinctive movements and postures that contrast strongly with the general fluidity of everyday activities.

The rigidity of the elements of animal communication gives animal displays their stilted, ritual-like nature and helps to keep them in a category of their own. When an animal is threatening a rival or courting a mate, the need to make a quick impact is crucial. There must be no doubt as to what is taking place. A process that assists in reducing ambiguity is the development of *typical intensities*.

A human example will clarify this. If I wish to contact a friend and the matter is urgent I can either telephone him or drive to his house. If I take the car, then the greater the urgency, the harder I press the accelerator and the faster the car travels. If I telephone him, the bell will ring at a set speed in his house, regardless of how desperate I am. All I can do to express the intensity of my desire to contact him is to let the phone ring longer and longer.

In this example, the action of pressing the accelerator produces a variable speed, while the action of dialing his number produces a signal with a fixed speed. Why should this be? Why can I not make the phone bell ring faster and faster to indicate my urgency? The answer is that the driving of the car is not a signal. My friend does not receive any information from it and its function is purely mechanical. But the telephone bell *is* a signal and it therefore requires a characteristic quality that makes it unambiguous. Its constant, fixed quality may be irritating to the desperate caller, but at least the friend knows that it is the telephone and not some other bell that is sounding. And if it rings and rings, he may eventually realise that the call is important and take the trouble to answer it.

This basic difference between the variable mechanical actions and the rigid signal actions is encountered time after time in the world of animals, but there is a slight difference. It is not easy for an animal to produce a response that is completely fixed and which has no variations at all. Only machines can do that. But it is possible for an animal to come close to this condition. Its display actions may not demonstrate fixed intensities, but they do show typical intensities. In such cases, there may be a little variation between a very low intensity response and a very high intensity response, but the difference is so small that it is of no significance.

An example from the courtship of birds will help to illustrate this. When the cut-throat finch male is courting his female, he raises his feathers in a particular way. This tells the female immediately that he is in a sexual mood and if she is interested in him she watches intently. If he sees that she is not moving away, the male then starts to sing his courtship song and the sexual encounter begins to develop. Feather-raising is the crucial first sign that a sexual interlude is imminent, and it is vital that it should not be confused with feather movements of a purely comfort nature, such as those connected with cleaning and temperature change. The male cannot risk adopting the feather-raised posture in a vague way. He must do it dramatically. He accomplishes this by erecting his feathers fully, into the characteristic ruffled posture that indicates sexual arousal. The switch is sudden, the feathers moving from sleeked to ruffled in a single step. In other words the courtship feather-posture now has a typical intensity, giving the courting male an unmistakable silhouette, regardless of the strength of his arousal.

Study of the strutting display of the male sage grouse has revealed a

Related species often fix on different forms of the same basic movement, as their displays evolve. The bowing action of the courting pigeon has become stylised in a number of different ways. Each species has its own special typical intensity that makes it distinct from the others. The blue-crowned pigeon tilts its head right down (1), the stock dove points its bill downwards (2), the plumed pigeon lowers itself almost as far and its strange crest tips forward slightly (3), the mourning dove holds itself upwards and inflates its neck (4), and the Luzon bleeding-heart rears back even farther and puffs up its chest to an extreme degree (5).

Examples of typical intensity are the crest-raising of the pink cockatoo (opposite), the inflation display of the sage grouse (below), and the head-throw display of the goldeneye duck (bottom).

similar situation. During the males' sexual advertisement to the females, a special posture is adopted and the bird's frontal air sacs are inflated and lifted twice in quick succession, with an accompanying explosive sound. Field observations have made it clear that the intensity of the display varies hardly at all, regardless of whether the males perform repeatedly to nearby females, or are far less aroused and display only occasionally to distant females. Despite obvious fluctuations in the strength of the sexual arousal of the males, they always gave the standard, stereotyped strutting display.

Signals showing a typical intensity of this kind are most often those that must be able to operate at long distances, where subtle nuances would have no value. For example, it is important that the female sage grouse recognise the male figures from far away and can see that they are performing the advertisement of their own species rather than that of some other bird. And the performers can still demonstrate their rising mood of sexual arousal by increasing the frequency of these displays while keeping the intensity fixed.

Also, signals of this type tend to operate most efficiently in dramatic contacts that are of a brief duration, such as when animals are threatening rivals, warning off predators, or courting their mates. The rigidity of the rapid signals increases the likelihood that they will be correctly interpreted, and again there is no value in subtle nuances – indeed, in such cases there is no time for subtlety.

Studies of the displays of goldeneye ducks have made it clear that almost all the signals of this species have typical intensities, there being only a few where there is a marked variability. Usually a statement like this would be based purely on observations made with the naked eye, but in this case the investigators filmed large numbers of displays and then carried out a careful analysis, measuring the exact timing of each display. To give two examples, the male goldeneye performs display actions called the head-throw and the masthead. In the head-throw the neck is stretched upwards and then thrown backwards until it touches the bird's rump. The bird then gives a little buzzing noise and makes two sideways flicks of the head, one to left and one to right, as he adopts his normal resting position. The duration of the display varies very little (from 1·08 to 1·5 seconds) regardless of the changing moods of the males. With the masthead display the male lowers his head to water level and holds it there for as long as three seconds. Then he stretches it out flat on the water and jerks it suddenly up into a vertical position, with the bill pointing to the sky. After only a tenth of a second, the male then snaps his head back to water level again, holds it there, and eventually returns it to its resting position. The main, central part of the display, the actual upthrusting of the head into its 'mast' position and its return to the water, is highly stereotyped, varying only between 0·55 and 0·75 of a second.

Typical intensities are also much in evidence in the world of fish. When fighting, rival fish go through a whole series of fluctuated moods, as the balance between fear and aggression shifts. A combatant that is becoming increasingly frightened and subordinate experiences a slow increase in its urge to flee and a corresponding slow decrease in its urge to attack. In theory it should therefore show a similar slow change in fighting posture from forward-facing aggression to the full retreat of fleeing, including all the angles in between. In reality this is not what fish-watchers observe. Instead they see the fighting fish adopt one of three basic positions: frontal display, broadside display, and retreat. These have become the three typical forms of the conflict, as it is outwardly expressed. This does not mean that the intermediate moods have been eliminated, even though the diagonal postures have become extremely rare. It simply means that the variable range of possible postures has been simplified during the course of

volution to a few simple, 'typical' forms that act now as distinct, separate ignals.

Some fish are slightly more complex than this. The Mexican swordtail as five fighting positions instead of the usual three. In the additional ostures, the fish tries to combine at one and the same time the frontal with he lateral display, or the retreat with the lateral display. So, its five typical lisplays are as follows: (1) frontal, facing towards the enemy; (2) s-shaped sigmoid) with the body broadside, but the head facing towards the enemy nd the tail away; (3) fully broadside-on to the enemy; (4) s-shaped with he body broadside, but the head facing away and the tail towards; and (5) acing away from the enemy. The whole range of varying motivations still xists, but it now finds expression in five 'key' stages, or *typical ompromises*, between all-out attack and all-out fleeing.

In this way, by the stereotyping of animal conflict behaviour, a whole epertoire of signals can be built up, making the body language of each pecies more effective. In the past observers have referred to these discrete isplay units as 'fixed motor patterns' or 'fixed action patterns'. We now now this is a slight exaggeration, but in biological contexts they are fixed nough. The displays have an immediately recognisable character and nake it possible for one species to set itself apart from its close relatives, ot merely in its colours and markings but also in its movements and ostures.

This species separation is illustrated well by the different kinds of tereotyping that occur in the bowing displays of pigeons and doves. In ach species the male performs a deep bow to the female when courting her nd this bow varies from case to case. To give just five examples: the stock ove bows as low as possible, showing off the bright patch of feathers on he back of the neck; the plumed pigeon reaches the climax of the bow with is bill pointing vertically downwards towards the ground, with the result hat his tall head-plume stands conspicuously erect; the blue-crowned igeon tucks his head down so that his bill touches his breast and his fan-ke crown is pointing to the ground; the mourning dove puffs out his neck,

In the breeding displays of the king penguins, as in those of so many birds, the intensity of the display of one individual is matched by another. Their similarity again reflects how widespread the phenomenon of typical intensity has become as a communication device.

displaying the iridescent patch there, and then holds his head stiffly up as though he is about to bow, but fails to do so; the Luzon bleeding-heart throws his head up and back, as if about to bow, but then holds it in the upper position, showing off the vivid 'blood-mark' on his breast. In each of these instances, the bow has become fixed at a special point of its arc, and this variation of the bowing movement has become emphasised by the addition of special display structures. The typical form of a movement and the colours and markings go hand-in-hand, creating the strikingly contrasting displays of the different species.

If all this emphasis on the rigidity of animal displays gives the impression that there is no subtlety in animal communication, this would be misleading. The subtlety is retained in other ways. A display that has become rather rigid in its form or its intensity can vary in the frequency with which it is performed. A display that has developed a typical posture can vary in the duration for which that posture is held. So the elements of shape, form, posture, frequency and duration of movements can each be rigid or variable, and this will differ from case to case. Each species can thus offer distinct, highly characteristic units of communication and, at the same time, show stronger or weaker versions of these displays.

When fish are fighting there are two common postures: the frontal confrontation and the broadside display. These two extremes are widespread and have typical intensities, but the intermediates between them are rare.

Facial Expressions

When animals engage in intense social encounters they usually do so face to face. Even if their bodies are twisted sideways, their faces are still directed at one another. The reason is obvious enough – they try to keep their eyes, ears and nose sharply tuned to the shifting mood of the companion or enemy. They also keep the biting mouth in readiness for a possible attack. Because of this, it is the facial region that comes under the closest scrutiny of any body area. The tiniest change of expression there acts as a tell-tale signal.

For many animals the possibilities for facial expressiveness are severely restricted. The muscles of the head region are rather limited in their actions. Reptiles and birds can do little but gape. Snakes cannot sneer, lizards laugh, parrots pout or sparrows snarl. The best they can manage is to open the beak menacingly as a warning that they are about to bite; or, with small nestlings, stretch the mouth open to its widest as an invitation to parent birds to provide food.

Only with mammals does facial expression reach any degree of subtlety. Even there it is only impressive in certain more advanced groups. The lower mammals have little more to offer than the birds. An opposum, for instance, has a lifelong, fixed expression that varies only when it gapes its mouth open in threat. But in three groups of mammals the face has become more mobile – in the hoofed mammals, the carnivores and especially the primates. They can send a variety of signals with their ears, their eyes and their mouths. And by combining different elements in different ways they can create a whole repertoire of faces, each corresponding to a special, complex mood.

These higher mammals are constantly reading one another's faces, checking for any small shift in friendliness, sexuality, fear or hostility. There is little individual variation. The facial expressions are common to all members of a species, develop naturally without particular learning, and remain comparatively unchanged throughout adult life. Reactions to them are also essentially inborn.

The majority of animal facial expressions are not limited to a single species. The frightened face of one sort of monkey is much the same as that of another. An alarmed horse looks much like a startled zebra. An aggressive dog looks the same as an aggressive jackal. But there are interesting exceptions – unusual expressions that are limited to just one, or only a few, species. Before considering those, what are the basic 'faces' shown by animals and what do they mean?

It helps to separate the three main organs of expression.

1 The Ears

- *Relaxed Ears* When nothing particular is happening the animal usually has its ears pointing sideways and slightly forwards. In this position they pick up sounds over a wide range. This is the 'neutral mode'.
- *Pricked Ears* When the animal hears a sound its ears become stiffly erect and rotate forward as the eyes focus on the source of the sound. This is the 'alert mode' in which the animal anticipates some kind of activity.
- *Twitched Ears* Animals with highly mobile ears may suddenly start flicking or twitching them. If these movements are comparatively slow and have pauses between each flick, the animal may be doing nothing more than turning the ears, like radar dishes, to pick up sounds coming from several sources at the same time. But if the twitching becomes rapid, then something special is happening. This signal indicates that the animal is in an acute state of conflict. The ears are responding by repeatedly readjusting themselves to suit first one mood and then the conflicting one. This is the 'agitated mode' and in some species it

For most animals, from moray eels to hippopotamuses, the only important facial expression is the gape. The mouth opens in a bite intention-movement that acts as a threat during hostile encounters. The facial region is too rigid to permit a more complex repertoire of expressions.

reflects a condition of barely controlled panic. With horses, it means that the animal may bolt at any moment. With some cats it has been developed into a highly conspicuous gesture, with the evolution of exaggerated ear-tufts. For example, when the caracal lynx twitches its ears, its hair-tufts flick through the air like fly-whisks. With their highly visible black hairs contrasting with the white hairs inside the ears, the tufts magnify every small motion of the ears and dramatically increase their signalling power during occasions of excitement or tension.

Elephants twitch their ears at times of high emotion, but the ears are so vast that the twitch is slowed down to a flapping action. In unemotional moments this flapping may be observed when elephants become overheated. The huge flat ears act as cooling devices for the giant pachyderms. There are many blood vessels near their surface and swishing the ears through the air helps to lower the temperature of the blood. At tense times, when the elephant is 'hot and cold' with emotion, its huge cooling system goes into automatic operation, becoming an important signaller of changing moods and of inner conflict. When elephants are aggressive, they modify the flapping slightly, holding the ears in the spread position longer than usual. This makes them look bigger and more intimidating to an enemy.

Monkeys and apes lack spectacular ear movements and signal little in this way, although some minor changes of ear position are possible under conditions of acute excitement.

- *Rotated Ears* When cats are becoming angry, they rotate their ears so that the backs of them face forward. In many species of cats, including the mighty tiger, there is a large white spot, ringed with vivid black, on the back of each ear. These display-markings are normally only visible from behind, but when a cat is made aggressive it twists its ears round and aims the spots straight at the opponent.

The ears are among the major organs of expression in the head region. When an elephant charges it raises its large ears, and this helps to make its already huge frame look even bigger. It flaps them back and forth while they are in this position, making them even more conspicuous. This flapping action is derived from a cooling response and occurs in attacks because the stress of the moment causes the elephant to feel overheated.

An alert cheetah pricks its ears and directs them forward, picking up any suspicious sound in front of it (above, left). But once it becomes hostile, it rotates its ears (above) in readiness for flattening them against the head should a fight ensue. All cats respond in the same way and the tiger (overleaf) has vivid white spots on the back of its head which are revealed when it rotates its ears in an aggressive moment.

The reason why the rotated ear is linked to an aggressive mood is that when animals are fighting they flatten their ears against their heads as a protection against injury, and before they can flatten them fully they must first turn them backwards. This means that, in terms of emotional condition, the animal with rotated ears is one that is halfway to a fight. So, when you see the backs of its ears you know you are facing a threatening, hostile individual. If you push the animal any further, its ears will go flat and then a serious fight may break out.

Flattened Ears This is the 'protective mode'. By pressing the ears hard against the sides of the head, the animal reduces the chances of them being torn by the teeth or claws of a rival. This is the ear posture of both the attacking individual and also the attacked. Once two animals are fighting they both show this reaction, regardless of which one was the aggressor and which the defender. It is an ear posture that is common to all higher mammals, whether horses, cats, dogs, or even monkeys. It is far less conspicuous in monkeys of course, but it can still be detected. As two monkeys fight, the skin around the sides of the head is pulled back and the ears shift very slightly backwards with it, a reminder of an era when their bigger-eared ancestors still had use of this particular ear protection movement. The only time when a fighting animal does not show the flattened ear posture is when it is totally dominant and fearless. Occasionally an attacker is so superior to its rival that it simply wades in, biting and striking without any apprehension, knowing from the body language of its enemy that there is no risk of retaliation. This is rare in the wild, however, where there is

nearly always some risk, even to the most dominant individual.

- *Airplane Ears* Some animals can be observed with their ears flopped out sideways from the head, with the openings of the ears facing downwards. This is a non-hostile signal and usually given by a submissive animal that is trying to appease its companions. Essentially it is the opposite of the Pricked Ears posture. There, the upright position of the ears is saying, 'I am alert and ready for anything.' Here, the Airplane Ears are saying, 'I have switched off. I am not bothering. You are in charge.'

  Both dogs and horses employ this ear posture, as a signal of inferiority. And they also use it during sexual encounters to say, 'I am not going to assert myself. You can approach me without fear of attack.' For this reason it is a signal that has sex appeal when a male and female are in reproductive condition and engaged in courtship. Occasionally, though, it can be misleading. Sometimes a female feels sexual enough to give the Airplane Ears signal to an approaching male, but then, when he comes close, she has second thoughts and repels him. When this happens to a sexually excited dog he shows signs of intense bewilderment and confusion, jumping back and forth agitatedly, not knowing what to do next. As he leaves the female she calms down, feels sexy again, and once more droops her ears in the airplane-wings position. As soon as he sees this, the male's sexual excitement mounts, and he comes close a second time. Again the female suddenly feels threatened and bites the unfortunate male, who squeals and bounds away, even more confused. To a casual human observer this looks remarkably like deliberate deceit on the part of the female and helps to explain the origin of the word 'bitchy', but the real explanation is simple enough. The Airplane Ears are indeed a sexual signal and attract the male, but the female, although aroused, is not yet sufficiently excited to permit the mating act to occur. She is in an ambivalent state typical of animals during courtship. She wants to attract the male but, having done so, wants to put him 'on hold' until she is ready.

Staring eyes often act as a threat in the animal world. This message is clear enough when we look at this lion, and yet we all too often forget it when we encounter our pets or companion animals. People stare intently at cats, dogs and other tame animals without realising that such a direct look can be intimidating to them.

2 The Eyes

- *Relaxed Eyes* When an animal is in a quiet mood, its eyes are open but not fully so.
- *Staring Eyes* This is the equivalent of Pricked Ears. When something has alerted the animal it opens its eyes wider, slightly increasing its range of vision in the process, and fixates the source of interest.
- *Blinking Eyes* An agitated animal increases its blink rate conspicuously, as if it is keeping its eye-surfaces ultra-clean, ready to see every tiny detail of what is about to unfold.
- *Frowning Eyes* In the typical frown, the eyebrows are lowered and the eyes are half closed. This is a protective expression, the brows being brought down to shield the all-important surfaces of the eyes. The eyes themselves are closed as much as they can be without blocking out the line of vision.
- *Glaring Eyes* Here the eyebrows are also lowered into a frown, but at the same time the eyes fight to keep as wide open as possible in an intense stare. This is a contradictory expression, with the animal trying to protect its eyes while maintaining as wide a range of vision as possible. As a result it is a difficult expression to hold for very long.
- *Closed Eyes* If an intensely active animal closes its eyes during an encounter it is attempting to switch off the incoming stimuli, and this acts as a signal indicating that the animal is subordinate and submissive. It has been dominated and has given up the struggle.

3 The Mouth

This is the most expressive part of the face. The reason why it can form a wide range of expressions is that it can move in four different ways. First, the lips can be pulled back horizontally, exposing the teeth to view, or moved forwards to conceal them. Second, the lips can be moved vertically, with the upper lip pulled up and the lower one down. Third, the mouth can be opened in a gape or shut tight. And fourth, the whole mouth region may be either soft and limp or hard and firm. In addition, the jaws may be opened and shut as a repeated movement and the tongue may be protruded. Together these variables make for a fascinatingly subtle set of oral expressions.

- *Relaxed Mouth* The resting posture, with the mouth lightly closed or slightly open, the lips almost or completely covering the teeth, but without any tension. A soft, neutral expression that becomes the 'baseline' against which other, more extreme expressions are judged.
- *Tight Mouth* The lips are pressed together and the short mouth-line is hard and rigid. This is the mouth of intense concentration. When used in social encounters, face to face, it indicates a dominant, passively hostile mood.
- *Gaping Mouth* The jaws are opened and the inside of the mouth is visible. This is the intention movement of biting and usually suggests that the animal may attack if pushed any further. It can however have other meanings. Accompanied by panting it can simply mean that the gaping animal is overheating. To discern the difference it is necessary to study the tension and shape of the lips. In heat-gaping the lips are pulled back and slack, and the tongue is prominent. In aggressive gaping the tongue is kept out of the way and the lips are more tense. Closer examination reveals that the position of the lips may be forward or retracted. In the forward position only the front teeth are exposed by the gape. In the retracted position all the teeth are visible. There is a key difference in mood between these two mouth postures. Briefly put, the more the mouth-corners pull back, the more scared the animal is. The more they are pushed forward, the more aggressive and fearless it is. If a dog retracts its lips vertically it means that it is dominant and hostile. If it retracts its lips horizontally, it means it is on the defensive and although prepared to threaten its rival, is also frightened of it. In many mammals this distinction is not made – a gaping mouth in a tense encounter simply means 'threat' without it being calibrated as more or less fearful. But the social life of dogs, cats and monkeys is so complex that these groups do need and do employ a whole range of subtle distinctions in facial expression.
- *Play Mouth* When an animal is in a playful mood it is important for it to signal that its attacks are not serious. It does this by using a special mouth posture: the mouth is open but the lips are stretched over the teeth to cover them completely, or at least as much as possible. Sometimes only the upper teeth can be covered in this way. The chimpanzee employs this expression when play-fighting, and it is sufficiently distinct from the threat-face for there to be no confusion. In dogs, the play-face is slightly different. The playful animal pulls its lips back as far as it can in the horizontal plane, but keeps them pulled over the teeth to conceal them almost completely. This has sometimes been called the canine 'grin' and it has been compared with human smiling.
- *Pouting Mouth* Among monkeys and apes there is a protruding-lips expression which makes the animals look as though they are offering their companions a kiss. It is in reality derived from the act of sucking at the nipple. The soft, relaxed lips are puckered and pushed forward as if seeking some object. It is an expression seen in friendly greetings and

When animals such as gibbons (above) and spider monkeys (below) yell or hoot, they have to open their mouths wide, and in so doing they expose their powerful teeth to view. This flashing of fangs is frightening to other individuals and the exposure of teeth is nearly always connected with aggression in some way. But this is only the beginning. In many species the exact way in which the teeth are shown indicates more precise mood.

The subtlety of canine facial expression is summed up in this encounter between two rival sledge dogs. Both are highly aggressive but there are small differences between the two that reveal that the one on the left is more domineering. The one on the right has lowered his head slightly and is sticking out his tongue in a special 'licking display'. This is a signal derived from juvenile behaviour, when pups lick the mouths of their parents in search of food morsels. In adults it is submissive, but this animal is clearly in a mixed mood because in other ways his face is still hostile. He is torn between fear and aggression. If he became more fearful, his ears would flatten and his mouth corners would pull farther back.

also at times when an animal is mildly miserable and needs some form of assistance or comfort. It signals helplessness and is sometimes used as a begging gesture.

- *Lipsmacking Mouth* This is another monkey speciality. The lips are opened and shut rapidly, making a gentle smacking noise, while the tongue may be thrust out of the mouth very slightly. This is observed as a friendly, slightly submissive greeting. It is derived from the actions of mutual grooming, when two monkeys sit down together and search through one another's fur for dirt particles and pieces of dead skin. Since grooming among monkeys is essentially an act of friendship, the Lipsmacking Mouth says in effect, 'I like you so much that I wish to groom you in this way.' During the course of evolution, the action has become exaggerated and speeded up, to enhance its role as an important signal. It now looks quite different from the true grooming actions, where there may be prolonged pauses between the eating of each discovered particle.
- *Teeth-chattering Mouth* If a friendly monkey becomes intensely frightened it cannot perform the lipsmacking action properly because its lips are fully retracted in the 'fear-gape'. But the friendly signal has been retained in a slightly modified form. The mouth still opens and shuts rapidly but, instead of the smacking noise, the sound heard now is that of teeth clattering together with rapid-fire action. This is the display of a very subordinate individual who nevertheless wishes to stay inside the group rather than go off to live the life of a social outcast.

Surprisingly, teeth-chattering is also found in young horses in precisely the same context, to signal inoffensiveness. With foals it is called 'teeth-clapping' and it is given to any adult horses that approach too closely. Unlike the monkeys, where the action may be shown by adults of any age providing they are in a strongly subordinate role, horses cease to show this behaviour by the time they are three years old. For them it is essentially a gesture of infancy.

- *Flehmen Mouth* Many mammals show the strange *flehmen* expression which accompanies intense sniffing. It is nearly always performed by males as a reaction to the urine of females on heat, but may occasionally be provoked by certain strong-smelling chemicals. The head is tilted up as the neck stretches forward. The top lip is curled upwards, exposing the upper teeth and sometimes even the upper gums. The mouth is slightly open and the animal appears to be momentarily lost in a kind of reverie, almost a trance. The impression

This is an aggressive snow leopard, with teeth exposed in a snarl, but with the mouth corners not retracted very far. The 'tall' snarl is always more dangerous than the 'long' one.

Hiding the teeth and pushing the lips forward in a pout is a non-aggressive expression in chimpanzees (above), as it is in humans, and is often accompanied by a greeting call. It is derived from the infantile expression in which the lips reach out to obtain something from the parents.

The special facial markings of the spectacled langur (right) help to highlight small changes in eye expression and mouth postures. The white mouth patch enhances the shape of the lips and the white eye-rings reveal more clearly the direction and degree of opening of the eyes.

This white-collared mangabey (below) has evolved a pair of vivid white eyebrow patches, markings that help to exaggerate slight changes in facial expression. As the animal's mood shifts during hostile encounters, its head skin is pulled up and back and then down and forward alternately. Increased aggression pulls it forward and increased fear pulls it back. This means that a monkey's eyebrows are rising and falling repeatedly during confrontations and the white markings highlight these changes.

given is that the animal is inhaling deeply and savouring the fragrance in the air. In human terms, it is reminiscent of a hungry man pausing to enjoy the delicious smell of cooking.

The monkey version of the expression is slightly different from that seen in other mammals. It is most vividly displayed in one particular species, the pig-tailed macaque from southern Asia, where it has been referred to as the Protruded-lips Face. The reason it has been given this name is that when the upper lip is raised and thrust forward, the lower lip follows it and presses up under it. There is little or no exposure of the teeth of the kind that occurs in hoofed animals and carnivores. The action does however occur in precisely the same context, the male macaque sniffing intensely at the rear end of the sexually active female, and it shows the same 'gazing-into-space' quality. There is little doubt that it is merely a primate variant of the basic response.

What is the function of this strange expression? Early authors suggested that the expression was a 'protection against unpleasant chemical stimuli'. They interpreted the upper-lip curling as an attempt to protect the sensitive nose from bombardment by over-pungent odours. The raising of the upper lip was thought to be an attempt to close the nasal openings and keep the nasty smell out. Why a sexually active male should wish to deny himself the rich fragrances of a female on heat, they failed to say. This explanation is clearly one designed by puritans to appeal to the fastidious and completely ignores the simple facts of mammalian courtship which, for most species, has a powerful olfactory element.

The true function of the *flehmen* mouth is precisely the opposite. It is a posture of the face that, far from keeping the sexual scent out of the nasal cavities, *traps* it there for prolonged savouring. Having adopted the expression, the displaying animal usually holds his breath for a moment, while continuing to thrust the top lip forwards and upwards. It is during this brief period that the animal, like an expert wine-taster, analyses the 'bouquet' of his companion's body, gleaning from this analysis her precise sexual condition. Because the *flehmen* face has such strong ties with sexual assessment, it has gone on to become a signal in its own right during the course of evolution and can now be given by one animal to another as a visual signal of sexual interest.

These are the main elements of facial expression for the higher mammals as they grimace their changing moods to one another. By combining them in various ways any horse, cat, dog or monkey can display a rich repertoire of signals to its friends or enemies, its mates or rivals, its offspring or parents. It can convey subtle changes in playfulness, sexuality, submissiveness, dominance, fear or aggression. It can signal its general level of activity, its alertness, agitation or calm. These are vital social messages that it must put across to members of its species and the mammalian face is wonderfully equipped and well positioned to transmit them.

In addition to the more common and widespread expressions there are also a number of specialised ones, restricted to a few, or a single, species. Some involve hair erection, as in the crested black ape from Celebes. This black-furred monkey boasts a conspicuous tuft of long hairs on top of its head which it is capable of raising and lowering as its mood fluctuates. When it becomes alarmed or fearful the crest is flattened, providing a highly visible indicator that operates even from a considerable distance.

The flattening of this crest reflects an underlying principle in facial movements: namely, that the more frightened an animal is the more its head-skin pulls backwards and the more assertive it is the more it moves forwards. This is an oversimplification, but it helps to explain many of the grimaces observed in animal encounters. Take the two extremes. The intensely frightened animal stretches its head-skin backwards with the result that (1) the eyes are pulled wide open for maximum range of vision, (2) the eyebrows are raised, exposing the patches of skin above the eyes, (3) the scalp skin is tightened, flattening the hair, and (4) the mouth corners are retracted. The intensely aggressive animal moves its head-skin forwards with the result that (1) the eyes narrow in a rigid stare as the enemy is fixated, (2) the eyebrows are lowered in a frown, (3) the scalp moves forward, erecting head hair and making the animal look larger, and (4) the mouth corners are pulled forward.

The Celebes black ape is not the only monkey species that has amplified and exaggerated this basic response. Certain mangabeys have evolved brightly coloured patches of skin above their eyes. These are vividly displayed when the animal becomes alarmed. As a further modification, mangabeys have developed an on—off flashing display of these colour patches as a friendly greeting signal. Most of the friendly greetings observed in the animal world (and humans are no exception to this) are based on signs of mild alarm. This is because meeting another animal nearly always involves some elements of apprehension, even it is a friendly contact. The on—off flashing of the eyebrows is a reflection of this slight unease.

Perhaps the most bizarre and dramatic of all animal facial expressions is the gelada baboon's lip-flip. This is the most extreme example of head-skin retraction known. When greeting another animal, the gelada pulls its scalp back like any other species and in the process exposes a pair of vivid eye-patches. But that is not all. As the mouth-skin retracts, the upper lip folds back on itself revealing the massive gums, the pale pink of their suddenly exposed surface contrasting strongly with the dark fur. The large expanse of naked skin displayed by the lip-flip is so odd to human eyes that the

The extraordinary 'lip-flip' expression of the gelada baboon from Ethiopia may look fearsome to human observers but to other geladas it is a friendly signal. It is in reality a dramatic pulling back of the lips from the teeth of the kind that goes with a less hostile mood. Aggressive geladas, like so many other expressive mammals, keep the mouth corners forward when attacking or threatening in a dominant way. Submissive or appeasing animals withdraw their lips in the horizontal plane – and when the gelada does this the upper lip automatically flips inside out.

In many mammals, when the male smells a female on heat he tries hard to increase the fragrance entering his nostrils by curling his top lip upwards. This curled-up face has been called the *flehmen* face, or flaring face. Here it can be seen in sexually aroused bighorn sheep (opposite, top), and it has even been observed in rhinoceroses (opposite, bottom). There is no hint of aggression in this expression, which usually heralds some kind of breeding encounter.

expression was christened 'the face that sank a thousand ships' by the first observer to encounter it. Strangely, although it looks so fearsome to us, with its exposure of huge canine teeth, it is in reality an extremely friendly gesture. If you happen to be another gelada you react to this display not with panic but with confidence. For it indicates, like other skin-retraction expressions, that the animal in question is more frightened than hostile. It is the gelada's unique way of saying to a companion, 'I wish you no harm.'

The huge, brightly-coloured mandrill has a facial expression that means much the same. But again, to human eyes, it looks anything but friendly. The mandrill keeps it jaws closed during this display but opens its lips in a curious way. The central part of its mouth is closed, while the mouth-corners are open. This gives the lips the shape of the figure 8 on its side. To us it looks like a savage snarl, but to another mandrill it says, 'I am friendly,' and it is most likely to lead to a grooming session or even play.

The origin of the mandrill 'snarl' is intriguing. It is the exact opposite of the most aggressive threat-face. Most species bring the mouth-corners forward and expose only the front teeth when performing a hostile threat. And they pull the mouth-corners back and display *all* the teeth as they become more submissive. The mandrill goes one better. When it is anti-hostile, it *closes* the front part of the mouth, shutting off completely the aggressive 'front-teeth' signal. At the same time, it gives the submissive 'rear-teeth' signal. A unique mouth-shape for a most unusual species.

Careful observations have revealed that the mandrill is also odd in another way. The big males yawn far more frequently than the females and it is thought that this gesture is employed as a mild threat display. All monkeys yawn when they are tired or experiencing a mild conflict, but the mandrill has apparently co-opted this action as an additional aggressive display and has increased its frequency accordingly.

The complex muscles that make all these facial expressions possible reach their peak not in dogs, cats or monkeys but in us. The human face is the most expressive in the animal kingdom. But it should be clear by now that the other mammals are not far behind.

Despite its appearance to human eyes, this mandrill 'snarl' is a friendly expression. The mouth corners are more fully opened than the front of the mouth. This is the opposite of the aggressive mouth.

Fighting Behaviour

Animals organise their aggression either territorially or hierarchically. In territorial defence, attacks are carried out on any same-species intruder, whether known personally or a stranger. Hierarchy disputes are always intensely personal, with known opponents fighting for higher places in the peck-order. Territory-owners frequently patrol their boundaries and perform 'ownership displays' – territorial birds sing, territorial mammals cry out their presence, as this roaring bull elephant seal (below) and bugling red deer stag (below, right) are doing, and territorial fish, such as these mudskippers (opposite, top), flash their fins at one another.

CONTRARY TO POPULAR OPINION, animals go to great lengths to avoid fighting. The phrase 'nature red in tooth and claw' has been widely misunderstood. Its only possible reference is to predatory behaviour, where the killer may well have blood-soaked weapons as it settles down to gorge on the carcass of its prey. But it has frequently been applied to fighting between rivals of the same species, with the assumption that, in the wild, animals are constantly duelling to the death to establish themselves in a dominant role. Nothing could be further from the truth. The overwhelming impression one gets from watching animal disputes is of remarkable restraint and self-control. The spilling of blood is not the norm – it is a rare event.

The reason for this restriction is not softness or kindness but purely selfishness. The fact is that if two rivals engage in a serious, physical fight, there is a strong chance that the winner of the bout will suffer some injury. The loser may be more seriously wounded or even killed, but that is no comfort to the winner if its own injury becomes infected and its health suffers. Even without infection, if an injury to a limb means that the winner cannot run as fast as usual then it may be unable to pursue prey, or escape from predators, on the following day. For both predator and prey, there is only a split-second difference between success and failure during the chase. Even a small wound on a leg may be enough to cause starvation in a predator or sudden death in a slowed-down prey.

As a result, animals will do almost anything to avoid actual physical contact with aggressive rivals. They use several devices to enable them to settle their disputes without going to this extreme. The first is to share out

the available space. This is done by the setting up of *territories*. These are defended spaces – home ranges on which the male territory-owner is more and more dominant as he approaches the centre of his personal area. At the borders, he is less sure of himself, as are his neighbours. They hardly dare to intrude on each other's territories and meet only on the periphery. There, their assertiveness is neatly balanced and the level of their aggression is reduced to a manageable level. They may display in a hostile way to one another, but the disputes do not become so intense that serious fighting explodes. In that it shares out the available space, the territorial system is rather benign and gives almost every individual a chance of reproductive success. Some territories may be a little smaller than others, or less well situated, but, apart from that, each owner gets a reasonable chance to build a nest, court a female, rear a litter, or simply find enough food for himself.

There is a simple way of observing the effects of territorial spacing, using a long, narrow aquarium tank. If two males of a territorial species, such as the ten-spined stickleback, are placed in the tank, and each is provided with a clump of aquatic vegetation that has been planted in shallow, moveable trays, it is possible to slide the territories they establish closer to one another or farther away. If at the start the two trays are placed at the far ends of the tank and the rival males are allowed to build their nests in the tangle of plants, one at each end, they will soon set up a boundary zone roughly halfway down the tank. There they will meet occasionally and threaten one another with aggressive displays but with little actual fighting. If the trays are then moved closer to the middle of the tank, each male will find himself on his own territory and also on his rival's territory at one and the same time. This they find intolerable and serious fighting breaks out almost immediately. Each male is much too close to the other one's nest-site and cannot retreat because he will be moving away from his

Detailed studies of chimpanzee colonies have revealed that the old idea of a fixed social hierarchy operating in a straight line from the dominant animal to the lowliest one is far too simple. In reality, these intelligent apes operate a much more complex society with subtle subdivisions and subgroups. Social gatherings may involve shifts of allegiance, bargaining and coalitions of specific individuals.

Many territory-owning mammals leave a personal scent on landmarks in their territories. Some do this by rubbing special scent glands on objects, others by defecating and still others by urinating. This klipspringer deposits scent on twigs and branches from a facial gland below the eye. With a network of these scent-marks it is possible for neighbouring territory-owners to know a great deal about their rivals' movements and condition without having to encounter them face to face.

own nest. The fish find it impossible to resolve this artificially designed dilemma and the fighting goes on and on in a completely unnatural way. If the trays are then slid back to their original positions, peace reigns once more, with just the occasional boundary encounter and the usual threat displays. This simple test demonstrates very clearly the value of the territorial system and the way it serves to reduce serious violence.

Territories vary from time to time and from place to place. Many animals only establish them in the breeding season, when they need to defend an area around their nest. At other times they move about together peacefully in large groups. Other animals only set up territories when the local feeding conditions are suitable. In areas where there is abundant food, they may settle down and space themselves out on defended patches, but if food is scarce they may become nomadic.

Another method of reducing the chances of serious fighting is the establishment of a *social hierarchy*. In the simplest form this involves a straight-line peck-order from the most dominant to the most subordinate member of the group. Each knows its place on the social ladder and refrains from challenging those individuals that are of higher rank. This reduces fighting to occasional clashes where relationships are either re-established or, once in a while as individuals grow older, altered.

In more recent studies of hierarchical groups it has emerged that this simple arrangement is often modified in some way, especially among the more intelligent species such as apes. In chimpanzee colonies it is possible to observe several important variations on the straight-line theme, including the establishment of coalitions to overthrow tyrants, collective leadership, dominance networks, divide-and-rule strategies, arbitration and bargaining. In short, the hierarchies of apes can be almost as complex and devious as those of human beings. But despite this, even with these remarkable animals, acts of physical violence are kept to a minimum. The rare outbursts quickly resolve themselves and there is always much more screaming and chasing than actual biting.

This restraint only breaks down seriously when conditions become excessively overcrowded. Then real and terrible violence may erupt and persist. It is because our own species has become so overcrowded, giving rise to a great deal of human bloodshed, that we tend to think that other animals must be as violent as we are. But the truth is that bloodletting is only typical of other species that have, for some unusual reason, become similarly overcrowded. This rarely happens in nature because all species

There is a considerable amount of sparring during the early phases of animal courtship and mating. One of the functions of the courtship rituals is to reduce the amount of hostility between the pair. This has clearly not yet been achieved between these lions.

have built-in population control mechanisms, limiting their numbers and preventing the breakdown of their social structures. Our giant brains with their advanced technologies appear to have inactivated that mechanism for our own species, a global development that we all may not live to regret.

Despite the general low level of animal aggression, there will always be those rare moments when two individuals do have a dispute to settle and it is worth taking a look at how this is done. What are their fighting techniques?

For many species the first contact made between rivals is at a great distance. Territorial males at the start of the breeding season roar, bellow, hoot, screech and sing their presence, announcing to all-comers that they have taken up residence on a particular patch. These distant threats establish the general area of the territory but refinements are needed. Specific boundaries have to be delineated. For this, face-to-face encounters are usually necessary although many mammals have adopted the device of depositing scent marks around the perimeter of their defended spaces. Scent-posts can deter possible intruders without the necessity for the immediate presence of the territory-owner himself. But even these usually require an occasional back-up encounter.

When rivals do meet one another near the borders of their territories there are three stages of intensity in the aggression that follows. The first is simply at the level of *threat display*. During such an interaction the most common phenomenon is the sudden increase in apparent size of the rivals. The animals use a variety of transformation displays to achieve this, erecting fins, spines, flaps, gills, wings, or other parts of their anatomy in such a way that they immediately seem much larger than when they first arrived. This growing in stature makes them more intimidating, on the principle that the bigger they are, the stronger they will be if driven to fight. The fact that the enlargement is more apparent than real does not seem to matter.

With fish, the threat enlargement is achieved by raising the gill-covers and spreading all the fins as wide as possible. Connected with this tactic, the gill-covers and the fins are often brightly patterned or coloured. With reptiles there are often erectile areas of skin, such as throat-flaps or neck-frills, and the animals try to make sure that these are directed towards the rival with maximum impact. With birds, all the body feathers may be raised, more than doubling the apparent size of the individual, and in

Because of their long necks, giraffes cannot engage in the frontal sparring typical of other hooved animals. Instead the bulls fight one another by 'necking', testing one another's strength by pushing, like human arm-wrestlers. Sideways swinging of the neck to strike mild blows with the blunt horns may develop, but these actions are rarely violent. There is a distinct peck-order among giraffes, established by the restrained fights, and really violent encounters only occur if a strange giraffe arrives in an occupied locality – an uncommon event.

Face-to-face fighting occurs whenever the threat displays have failed to settle disputes. But even at this stage of actual contact, the attacks are stylised, with a great deal of head-to-head pushing, butting and shoving, rather than all-out onslaughts. Ritual fighting is seen here in bighorn sheep (left), rhinoceroses (opposite, top left), wapiti (opposite, top right), gemsbok (opposite, bottom) and hippopotamuses (previous page).

addition crests may be erected and wings and tails spread wide. With mammals, bristling hair has a similar effect.

As the threats continue there is frequently a great deal of mouth-gaping, as a way of signalling readiness to bite the opponent. If feet are likely to be used in real fighting, there may be much pawing the ground or swiping the air. While this is going on the animals may be facing one another in tense postures, or they may start to circle around one another suspiciously, each preparing itself for the first lunge of the other. With aggressive noises, enlargement displays and incipient attack movements, the rivals may continue to intimidate one another for some time. Eventually, either one of them will slink away, conceding defeat, or they will have to give up their display fight and switch to the real thing. But even when this happens, there is still one more stage to go before serious violence erupts. This is because the most usual type of fighting observed in the wild is what might be called Ritual Fighting or Half-fighting.

These are inhibited contests in which the opponents pull their punches.

Boxing matches occur between kangaroos. The strategy is to get into a position where, with the support of the powerful tail, it is possible to deliver a vicious blow with the huge hind feet.

Bengal monitor lizards fight like sumo wrestlers, each grappling with the other's body and trying to throw it over. The large reptiles are capable of rearing up on their hind legs as a size-increasing threat display before they move in for the fight.

Each may be capable of doing terrible damage to its rival, but since this would only provoke similar actions in retaliation they both limit the violence of their actions. Real contact is made and these bouts are indeed physical contests, but the kinds of attacking actions used are rather stylised or restricted. In many cases special weapons have evolved that enable two males to engage in ritual combat of a dramatic kind, but without much risk to their important or sensitive organs such as the eyes.

In the spring the sound of horn crashing on to horn, or antler on to antler, may be awe-inspiring, but in reality all that the contestants suffer at the end of a long bout of sparring is a slight headache. Just occasionally there may be a tragedy, with internal organs suffering rupture. There are also instances when two pairs of antlers become irreversibly locked together and the two animals cannot separate themselves, no matter how hard or how long they struggle. In those rare cases both contestants die, but the normal outcome of a deer-fight is the final retreat of one stag while the other stands triumphant on his territory. With warthogs the risks of fighting have also been reduced. Their curved tusks must originally have been a serious danger to their eyes. However, during the course of evolution large fat-pads have developed, two below the eyes and two on the cheeks. These are the warthogs 'warts' and they act as barriers, protecting the sensitive parts of the head and making ritual battles safer.

A number of animals use butting instead of the more damaging biting, as a basic fighting technique. Once again they can hurt the opponent without causing serious injury. Marine iguanas on the Galapagos Islands have evolved armoured 'crash-helmets' on their heads with which they ram one another. Bitterling fish also develop horny warts on their heads during the spring and the males then butt one another in their territorial disputes. Other fish employ water jets as weapons. By beating their tails with great vigour, they can send powerful currents of water towards one another's sensitive lateral-line organs – the balancing organs of fish – and inflict pain in this way without actually ripping off scales with their teeth, an action that could quickly lead to infection. Giraffes employ their long necks for curiously inefficient bouts of side-swinging, in which each animal tries to ram the other one's head. Bearing in mind that the legs and feet of giraffes are strong enough to kick a lion to death, this neck-fighting is clearly a safer way of fighting.

If a ritual fight cannot settle a dispute then, ultimately, there has to be an all-out assault and blood will flow. This is the final phase of animal aggression. In extremis, the claws are out and fangs are flashing. Teeth are biting, ripping and tearing. No holds are barred. On the rare occasion when this happens the battle is usually a quick one. Unless the circumstances are exceptional, the entire fight will probably take no more than a few seconds. The loser is quickly vanquished and runs for its life. It will hardly ever be pursued because the goal of animal fighting is winning, not the destruction of the enemy. An animal that has been defeated is of no further interest to the victor.

Animal fights organised by humans are a different matter. Dog fights between highly trained pit bull terriers may last up to two and a half hours, until the animals are mutilated and totally exhausted, but this tells us more about human beings than it does about the true nature of animal aggression. The dogs concerned are put through elaborate and painstaking training sequences and the conditions under which they fight are designed to maximise the persistence of their attacks on one another. These are no longer normal dogs and they are a thousand times more violent than their supposedly savage wild ancestor, the wolf. But they do serve to remind us of just how far normal aggression can be magnified under special conditions – a salutory lesson for the human species.

When avian fighting erupts in water there are two techniques for defeating the enemy. The coot (above) uses the peck-and-kick method, tearing at the rival with long claws and aiming pecks at the head. The goose, in this case the Canada goose (top), employs the grab-and-beat technique, grasping the opponent's neck or breast in its powerful beak and then beating at it with the hard leading-edge of its wing.

Submissive Behaviour

One of the reasons why animal fighting is so restricted is that weaker animals have a variety of techniques for switching off the hostility of their stronger opponents. By performing special submissive displays they are able to transmit a clear message that says, 'You are the winner. I give up.' It is extremely rare for an attacker to ignore these signals, because once it has won the contest there is no point in risking possible injury during a further onslaught that might cause the loser, in desperation, to make one final retaliation.

There is one golden rule for a beaten animal and that is to make its submissive display as different as it can from the threat display of its particular species. If possible it should be the exact opposite in as many ways as its body will allow.

An example will clarify this. When ten-spined sticklebacks fight, each male threatens with (a) his head lowered, (b) his skin darkened to a jet black, (c) his white ventral spines raised, (d) his dorsal spines lowered, and (e) his tail fin spread. When one of the males has been beaten he adopts a submissive posture in which (a) his head is raised, (b) his skin is pale and cryptically marked, (c) his ventral spines are lowered, (d) his dorsal spines are erect, and (e) his tail fin is limply closed. In other words, he is in every respect the antithesis of the hostile fish. In this way he leaves no doubt about his abject, beaten condition.

This same head-down for threat and tail-down for submission is seen in a number of other fish species, including gouramis, swordtails, danios and some cichlids. Why these particular postures should be so widespread is not known. Opposites have also been observed in other species. The firemouth cichlid employs a dorsal roll as a submissive act. In this fish the threat colours are largely on the underside, and there are some on the flanks. When a female of a pair was seen swimming past her rather aggressive male, she rolled over on to her side as she came close to him, showing him only her upper surface. As soon as she had passed him, she rolled back into her usual vertical posture again. In her moment of submissiveness she had effectively hidden both her ventral and her lateral

Two jackals meet (above). The one on the left stares threateningly and the one on the right looks away. Averting the gaze is an act of inferiority and signals this animal's weaker status.

The jackal in the centre of this picture (right) crouches in a submissive posture, with its neck lowered, and its hind legs bent, its tail tucked tightly between them.

Submissive wolves try to make themselves as small as possible in relation to the dominant animal. Two extreme degrees of lowering are a lying-down posture (left) – the dominant animal towers above – and complete out-and-out submission in which the subordinate lies upside-down on the ground (below).

threat markings and in this way had reduced her stimulation of the male's aggression to a minimum.

With river bullheads, the threat display consists of a darkened, almost blackened head, spread fins, raised gill-covers, opened mouth and the body raised off the bottom by the ventral fins. When behaving submissively, the bullhead is pale and cryptic, with all fins folded, the gill-covers lowered, the mouth closed and the whole body pressed flat against the substratum. Again the two displays could not be more different.

Similar opposites occur in many species, but the details are not always the same. In the Canadian rock bass, for instance, the threat posture, unusually, involves a flattening of the pelvic fins against the body, and the colour then is black. When these fish are behaving submissively, they raise these fins which then appear intensely white.

Submissive displays are also common in reptiles. Lizards manage to switch off the aggression of their companions by trampling or 'paddling' with their front feet. The legs patter up and down and the head nods, as if the animal is running away, but it stays rooted to the spot. This 'running-on-the-spot' response is clearly derived from an urge to flee that cannot find expression in a proper retreat. It serves to inhibit the attacks of dominant companions and may be used simply as a reaction to the approach of a dominant animal, even when no actual threat is offered.

In birds and mammals, the most popular submissive actions are those in which the body of the animal is made to look smaller, flatter and lower. It is also characteristically motionless. In this cringing, crouching condition it looks the opposite of the erect, puffed up, enlarged dominant individual. This contrast is not unknown in human societies and it is something we share with our closest relatives, the chimpanzees.

Subordinate chimpanzees greet their superiors in a servile manner whenever they approach them, or when they themselves are approached. They emit a soft grunting noise, like heavy panting, and lower their bodies far enough for them to be able to *look up* at their dominant companions.

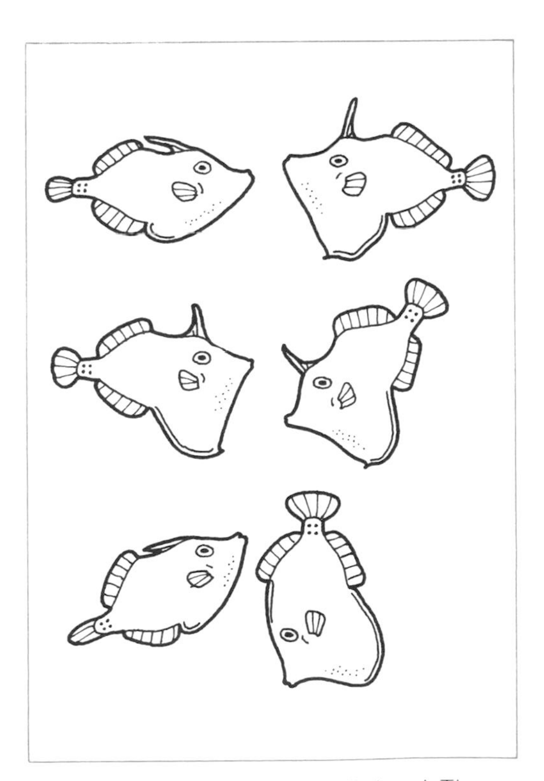

The aggressive file fish lowers its head. The submissive fish raises its head. The aggressive one spreads its fins. The submissive one closes its fins. The submissive display sends signals opposite to those of the threat display.

The European white stork on the right has just landed at its nest and greets its mate with a submissive clappering display. In this, it opens and shuts its bill noisily and throws its head right back until it touches its body. This is an exaggerated intention movement of take-off and its signal is one of non-hostility. It says, in effect, 'I mean you no harm, and is a way of calming the mate following a period of absence.

When the dominant animal is already sitting down, this is not easy and requires truly debased grovelling. In addition the weaker animal makes a series of quick bows, each dip of the body following so closely on the last one that the movement has been called 'bobbing'. Occasionally the subordinate brings to the dominant a small object such as a leaf or a stick and, to complete the servilities, it may also kiss the dominant one's feet. Bearing in mind that in human society many of these gestures were once part of church ritual, it is perhaps not surprising that bishops have been referred to as primates.

These displays are most commonly seen in submissive male chimpanzees. Females prefer a more sexual approach, usually offering their rumps to dominant animals for a token sniffing and perhaps mounting. This is one of the re-motivating actions that suppresses aggression by arousing an alternative response – juvenile behaviour, sexual behaviour and grooming are all commonly employed in a submissive context. Cut-off actions, with closed eyes or hidden faces, are also widespread. Together all these submissive displays serve to appease the wrath of the victors of fights, and the overlords of animal society. They are important because they help to reduce the length of disputes, to terminate fights and to enable the weaker members of society to remain in close proximity to the stronger without arousing repeated hostility. Without them, closely-knit, complex animal societies would not be possible.

Animals that exhibit colour changes with breeding condition, such as these tilapia fish, may quickly lose their breeding colours as a display of subordination. In this encounter the dominant fish is on the right. The weaker fish is already beginning to turn darker as it loses its bright reproductive colour tones.

Courtship Displays

At the start of the breeding season animals perform their most extraordinary displays. They leap, dance, twist and turn. They ruffle their feathers, bristle their hair, erect their spines and puff up their pouches. Their previously dull bodies suddenly flash and glisten with bright colours and vivid contrasts. They strut and weave, nod and twitch, and generally behave with such fire and intensity that the animal-watcher can only marvel at the richness of the scenes unfolding before him.

For centuries the 'love-play' of animals has fascinated observers but until recently there was little attempt to analyse what was taking place or to try to understand it. The strange actions were simply recorded as 'comical antics' and left at that. The reason that words like 'play' and 'comical' were used was that the courtship actions frequently seemed to be so exaggerated and uneconomical that the animals did not appear to be treating the problem of breeding in a serious or workmanlike fashion. But nothing could have been further from the truth. Even the most bizarre courtship ritual is strictly functional and entirely necessary. If the actions performed and the postures adopted appear rather odd and unexpected, this is no accident. Their oddity and the element of surprise are an integral part of the communication system that is in operation. For it is a system that involves the trapping of the attention of the sexual partner, and nothing catches the eye quite so well as an abandoned flashing of bright colours, or sudden, noisy movements. Courtship is a drama, not a comedy, and its dramatic displays all have a special meaning. It is just a matter of knowing what to look for.

The easiest way to understand animal courtship is to consider the problems faced by the male and female before they can achieve a successful mating. First, each animal must *find* a potential mate; second, it must *identify* it as being of the correct species, gender and age for mating; third, it must *attract* it, so that close proximity is achieved; fourth, it must *arouse* it so that it is physiologically ready for mating; and fifth, it must ensure that the arousal level of both itself and its mate are perfectly *synchronised*. Only then can successful copulation take place.

For some species these five goals are easily reached, and the courtship is simple or almost non-existent. Many lower forms of life react purely to certain environmental conditions of light and heat combined with some chemical signals from the mate. They come together and shed their eggs and sperm, and immediately go their separate ways. In such cases there are no displays or dances and no elaborate rituals to observe. But higher up the evolutionary scale social life becomes more complex and difficulties arise. The first concerns the location of the mate.

Animals that live in a closely knit group throughout the year have no trouble in finding a mate, but many species move back and forth between winter feeding grounds and summer breeding grounds. When they reassemble each spring there is the immediate problem of locating or relocating a mate. Bird species that form pairs may be capable of carrying a memory of a specific nest-site and returning there with amazing precision after journeys covering thousands of miles. But for many other species, and for all young animals facing their first breeding season, there is the need to transmit some kind of advertisement signal to announce their presence on the newly occupied breeding grounds.

These advertisement signals may be smells, sounds or sights. The sensitivity of certain insects to the courtship scents of their species is almost beyond belief. A female silk moth is capable of releasing into the air a pheromone (a chemical signal) called bombykol that can be detected by males up to 7 miles away. When she is in the mood to mate she flutters her wings, scattering the scent produced by special glands on her abdomen. It drifts through the air in unimaginably small quantities until tiny particles

The courtship signals of female atlas moths are sexual odours transmitted through the air over great distances. The elaborate male antennae are capable of detecting the female signals from several miles away. In some species males have been shown to be capable of responding to as little as one molecule of their female's scent, and to be able to detect it 7 miles from the female.

of it come to rest on the antennae of far distant males. These antennae, like ornate television aerials, are ever alert to incoming fragrances, checking them and analysing them until precisely the right one is received. Then the males fly off in the direction from which it arrived. As they get closer to the source, the minute traces of bombykol become slightly stronger, and this enables them to home in on the chemically displaying female.

The detection abilities of these and other insects seemed so incredible that it was tempting to suggest that they were using some kind of as yet unknown sense organ to locate one another. But careful tests proved beyond doubt that this was not the case and that smell was all that was involved. Synthetic bombykol was produced in the laboratory and was used to test the reactions of male antennae. Astonishingly, it was found that the receptor cells on the antennae were capable of identifying and responding to as little as a single *molecule* of the scent. This level of sensitivity to a landscape of fragrances is almost inconceivable to us, with our crudely inefficient noses. It would be easier to explain a rainbow to a blind man than to convey an impression of what it must be like to live in such a world.

It is much easier for us to understand the sexual advertisements of birds, where sound plays such an important role. The migratory males, arriving slightly before the females, set up individual territories and then start to sing their loud songs, time after time, announcing their whereabouts to potential mates. Human ears, unlike human noses, are sensitive enough to enable us to distinguish all the different song-patterns and to understand the way in which listening females can recognise the particular song of their own species. The presence of many different kinds of birds in each location has led to the development of more and more complex song-patterns, as each one keeps itself distinct from the others and easily identifiable.

Birds are not the only songsters, although they are certainly the most

The calling of frogs varies between a deep croak, a cricket-like chirrup or buzz, and a high-pitched chiming, according to the structure and resonance of their inflated air sacs. When a frog calls it keeps its mouth tightly shut and creates its sound by driving the air in its vocal sacs back and forth between its mouth and its lungs. Each species has its own distinctive call.

The humpback whale sings a song that can be heard hundreds of miles away underwater. It is the longest and most complex song known in the entire animal world, each performance lasting up to 30 minutes.

The singing of male birds during the breeding season advertises their presence to prospective mates and to territorial rivals simultaneously. Many courtship displays have this double function. The songster here is a great reed warbler.

An unlikely songster is the grasshopper mouse. This carnivorous American rodent has an elaborate courtship ritual involving a great deal of bipedal standing and the singing of a strange chirping song. The result of the courtship is that, after much chasing and nose-rubbing, the pair establish a tight bond of attachment and share fully the parental duties of rearing their young.

melodic. The 'singing' of chirping insects, croaking frogs and bellowing mammals also fills the air in the breeding season. Even fish of some species manage to produce watery grunts and clicks. The so-called silent depths of the ocean are, in fact, far from silent, but the noisiest of the sub-aquatic songsters is not a fish but the gigantic humpback whale. This species sings the longest and most complex songs in the animal world. A single song is never shorter than 6 minutes and may last for more than 30. And almost as soon as the song is ended, it starts again. Each song is made up of a sequence of themes and each theme comprises a number of phrases. It seems that the reason why humpback whales in the breeding season sing for hours on end is that the quality of the transmission of sound through the ocean waters varies considerably. At bad times, the song can only be heard a few miles away, but at good times it is audible, amazingly, at a distance of hundreds of miles. In an environment of limited visibility, this provides these huge animals with a remarkably efficient communication system and perhaps the most impressive long-distance advertising outside human television.

By comparison, all visual displays are strictly limited. They can only operate over rather short distances, but for predominantly visual animals such as birds they nevertheless play an important part in the run-up to the mating act. Many species adopt intensely bright plumage in the moult just before the breeding season and then station themselves at vantage points where their vivid colours can be seen by other members of their species.

All these displays put the performers at risk. Unless they happen to belong to an invincible species, their advertisement rituals can attract dangerous predators as well as potential mates. Some predators thrive by exploiting the need of animals to announce their sexual presence. Certain kinds of bats tune in to the croaks of breeding frogs and swoop out of the darkness to pick them off in mid-song. Snapping turtle operate a similar system and in the breeding season the pickings are rich.

A delicate balance is needed. If animals display too much, they die; and if they display too little and fail to find a mate, their species dies. The amount of advertising must be restricted according to the vulnerability of the particular species: whales can sing longer than dolphins, for example, and poisonous frogs can croak louder and longer than harmless ones.

Having risked life and limb to meet a potential mate, the first hurdle is over and the prospective partners must now confront the second: identification. Before the courtship proper can begin, each animal must ensure that it is confronting a member of its own species. In some species the long-distance advertisement signals also operate as close-quarters identification devices, but in others additional species-labelling is required.

Where large numbers of different species inhabit the same general environment, each of them must clearly label itself with its own 'species flag'. Just as human flags and national colours are most in evidence when large groups come together at competitive gatherings such as the Olympic Games, so too are animal flags most seen in high density settings, where a rich source of food supports many species in close proximity. The more species there are, the more complicated the markings become. The hundreds of fish species that inhabit the lush world of the coral reef have become the most intricately marked and most brightly coloured of any living creatures. The many kinds of waterfowl that gather on lakes show a similar ornateness and intensity of coloration.

The same is true of grassfinches on grasslands, as one example will demonstrate. The male zebra finch is essentially a grey bird, but with no fewer than ten distinct colour patches on his body. He has (1) a white rump patch, (2) chestnut ear-patches, (3) a white belly, (4) a red bill, (5) orange legs, (6) white-spotted chestnut flank feathers, (7) a black breast bar, (8) a black-and-white finely barred throat, (9) a black-and-white banded tail, and (10) a black-rimmed white cheek.

The zebra finch is found all over Australia, where it overlaps with twenty-one closely related species of grassfinches. Among these there are thirty instances of individual markings of the zebra finch appearing in the same form on other species, and an additional sixty cases where, although the individual markings are not exactly the same, they are similar. In other words, what seems a perversely and unnecessarily complicated set of colour markings on the zebra finch, proves on closer inspection to be essential for precise species identification.

Significantly, the race of zebra finches that lives on the island of Timor, where there are far fewer grassfinch species present, has simpler markings.

In addition to species identification there is also the problem of telling males from females and adults from young. This may be done by sounds, smells or movements, but in many cases it is also based on different visual displays. Sometimes the differences are small, as in the familiar budgerigar. Here the sex difference lies in the colour of a small patch of skin just above the bill, called the cere. Male budgerigars have a bright blue cere, females a dull brown one. If males have their ceres painted brown they are courted by other males. If females have their ceres painted blue, they immediately lose their sex appeal and are treated as rival males.

A similar situation exists in the red-shafted flicker, an American woodpecker, in which the visible difference between males and females is also a small colour patch on the head. In this case, however, it is a red 'moustache' at the corner of the male's bill that distinguishes him from the female. When females were caught, given false moustaches and released, they were immediately attacked as if they were males. Astonishingly this applied even within a mated pair. A mustachioed female was attacked by her own mate, revealing the overwhelming importance of this small gender signal.

The Australian zebra finch has complex markings in the male that at first sight appear to be unnecessarily elaborate. But it lives in a region where there are twenty-one other closely related species, all with complex markings, and only its unique combination of patches of colour can isolate it as a distinct species. (The male is on the right, his mate on the left.)

Many species have juvenile signals that are just as powerful. In the zebra finch, the juvenile bill colour is black, and this is the only visible difference between the fully grown juvenile and the adult female, whose bill is red. If red nail-polish is applied to the beak of a juvenile zebra finch it is instantly courted by adult males and even by its own father. Again this single gender signal overrules all other clues concerning the identity of the bird. As soon as the red colour is taken off with some nail-polish remover, the 'anti-sex' signal of the black bill comes into operation again and the young bird is left in peace.

The simplicity and power of these reponses seem strange to us. No husband would fail to recognise his wife simply because she appeared wearing a false moustache; no father would fail to recognise his daughter simply because she put on red lipstick. Yet for birds, certain key signals obliterate all other considerations. The important lesson here is that if we are to understand animals we must try to see the world from their point of view. Hard as it may be, we must stop looking upon their brightly coloured feathers as 'pretty' and accept that each marking, each patch, plays a vital role in their social lives.

One question that has given rise to a great deal of debate is why it is nearly always the male birds that are brightly coloured while the females are usually dowdy. For many species it is because the males are less important than the females during the breeding season. It is a risky business being brightly coloured because it attracts predators. The females, with heavier breeding duties than the males, cannot afford to take such risks.

Where species are polygamous this makes sense. The brightly coloured males display to the inconspicuous, heavily camouflaged females and try to attract as many as possible for mating. The females then leave to nest, incubate and rear the brood on their own. If one of the males is killed by a predator his place can quickly be taken by one of his rivals and no females need go without a mating. This is obvious enough, but the debate arises over the question of why monogamous bird species should similarly show different male and female forms, or sexual dimorphism. If the male and female share the parental duties and are linked together by a tight pair-bond, where is the advantage in making the males more showy and therefore more expendable? In such circumstances one would expect to find the two sexes equally camouflaged, but this is not always the case. There is no clear answer to this, but there are several clues.

Coral reef fish, such as butterfly fish (above) and triggerfish (below), display incredibly complex colour patterns but these are necessary for species identification among all the other rich colours and patterns of the reef, where many species live close together. This particular triggerfish rejoices in the name of Picasso fish, or *humu-humu-nuku-nuku-a-puaa*.

Obviously the female is vital as it is she who lays the eggs. The male cannot do that for her, although he can help her build the nest, incubate the eggs, feed the nestlings and defend the nest-site. If the female is incapable of doing any one of these tasks by herself, then the continued presence of the male is also vital, But if, despite considerable difficulty, a 'widowed' female could do it all by herself, then her role remains the more important one, even though under normal circumstances there would be a great deal of sharing – in such cases the male could still be brightly coloured and sexual dimorphism would still work.

Where the male duties are essential, however, there are still some cases of sexual dimorphism. One explanation for these is that they relate not to breeding activities but to feeding. For example, when males are bigger, with stronger bills, there may be a degree of food specialisation that reduces the food competition between the mated pair. This is a less obvious advantage but it seems to exist in a number of cases.

Another question that arises is: why be brightly coloured at all? If it exposes the males to danger, why do they not stay camouflaged like the

Polygamous birds, such as peafowl (right), show extremes of dimorphism, with the males displaying wildly exaggerated plumage to the female. Here, where the display of each male competes directly with the displays of his rivals for every female and every mating, the pressure is on to produce the most extreme and most eye-catching form of colour pattern. The female is dowdy compared with the male but she cannot risk being conspicuous to predators as she is about to nest and rear her young on her own.

Many species display special gender signals that separate the males from the females at a glance. In the red-shafted flicker the difference is a small red 'moustache' on the male (below) that is missing in the female (bottom). Tests have revealed that if a 'moustache' is painted on a female flicker she is treated like a rival male.

females? The answer appears to be that bright display colours provide a courtship short-cut. Dull males must excite their females sexually purely by their courtship dances, movements and postures. This can be time-consuming. If they can show off a bold patch of colour that has a unique sexual quality, they can initiate arousal more efficiently.

For the female it is important that she should pick a robust, healthy male as the father of her offspring, and one of the surest signs of health in a bird is the sleek condition of its plumage. So super-sleek feathers are at a premium. Any developments that make the feathers shimmer and shine are therefore attractive. And shimmering and shining can readily be emphasised and exaggerated by the addition of bright colours. Once this association has become established it can go on and on, until we reach the unlikely extremes of the peacock's tail – the ultimate bird display. Most species cannot afford to go that far. The displays would become too cumbersome and the males would fall easy prey to hungry hunters, but peacocks are driven on by the special circumstances of their courtship performances. Typically, in the wild, a group of males displays together and the female mates with the one that traps her attention most effectively. Under these conditions it is easy to see how more and more spectacular tail patterns could have developed as male competed with male. The cost in predation, however, is known to be heavy. The peacock seems to have pushed the display system as far as it can go.

Visual courtship displays are most dramatically expressed by fish, reptiles and birds. Mammals rely more upon smells and are consequently far less colourful as a group, although there are some notable exceptions, especially among the primates. But regardless of which sense organs have been dominant in attracting the sexes to one another, there now arises the next major stage of the courtship sequence, that of arousal. And it is not merely a case of one sex becoming aroused, but of both becoming excited together. Synchronisation of arousal is the primary function of a great deal of the courtship ritual that follows the initial attraction and coming together. It accounts for the complexity and strangeness of many of the courtship dances and ceremonies.

The problem with attaining mutual arousal is that one sex is nearly always slightly ahead of the other. Before mating can take place the less excited partner must be brought up to the higher level and this usually requires frequent repetition of the courtship sequence. It is this repetition that gives it its ceremonial quality. Some examples will help to explain this.

The Mexican swordtail is a small tropical fish in which the male possesses a dramatically elongated tail fin – the so-called sword. It is soft, however, and is of no aggressive value. Its function is that of a visual stimulator that arouses the female's sexual interest. The problem for the male is that if he is to excite the female with it he must somehow shiver it close to her face so that she receives the full impact of its 'maleness'. But when he swims up to her she becomes frightened and flees before she has had a chance to respond to his sexual display. The courtship dance of the swordtail therefore develops a tango-like quality, with the male repeatedly swooping up to and around the female in such a way that she has no direction in which to turn without getting yet another eyeful of his flashing sword.

The dance technique of the male swordtail involves several distinct 'steps'. The first is the 'pass' in which he swims up behind the female, passes her and then brakes suddenly. He is now just in front of her, so she cannot flee forwards as she appears inclined to do. Instead she too must halt, and in that moment he goes into the second movement and backs up to come alongside her. This is the position he requires for mating, but she is not yet aroused enough and so she turns away from him. As soon as she starts to make her move he makes his third step, swooping around her front end in a U-curve and wrapping his coloured sword around her face. Now she cannot turn left or right. He envelops her and she must stay still,

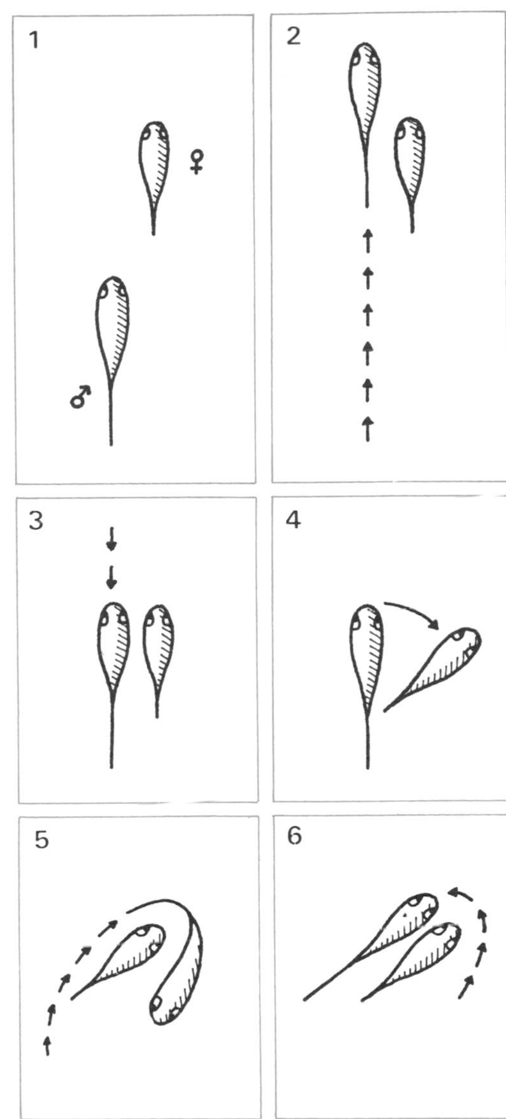

The Mexican swordtail dances to his female by quivering his sword in front of her face. To do this he has evolved a set of special dance movements that block the female's escape. In this way he can repeatedly stimulate her until she is ready to mate.

The males of many species perform ritualised movements, adopt special courting postures and carry out conspicuous dance routines as ways of arousing their mates to a point where copulation can take place. The female may be too frightened or too aggressive at first, and repeated displaying is necessary to suppress these non-sexual feelings and arouse her to a point where she is receptive. Many males achieve this by undergoing a 'transformation display'. Here the male rifle bird spreads his wings in an amazing arc that frames his yellow gape.

If one part of an animal's body is brightly coloured, this nearly always means that at least one of its displays is accentuated by this marking. The courtship dance of the blue-footed booby, as might be expected, involves a little tap-dance routine that is made more vivid by the unusual coloration.

The three-spined stickleback male develops bright colours in the spring, builds a nest and then waits for a female to arrive. When she does, he performs a rapid zigzag dance to her, moving first as if to attack her (below), then halting and darting towards the nest, then back towards her, then away again. If she is ready to spawn she lowers her tail and follows him as he zigzags towards the nest entrance. This sequence arouses the female, guides her, and synchronises the arousal of male and female.

at least for a moment. As she hesitates, he slowly backs round her head to come alongside once more.

This dance may be continued for ages, until the female either escapes or becomes more and more aroused by the male's sexual signals, to the point where she will stay still and allow him to copulate with her.

The essential feature of this courtship is that it simultaneously reduces fear and arouses sexual feelings in the female. One of the main problems that prevents males and females from mating as soon as they have come together in the breeding season is that their moods are not purely sexual. As part of their reproductive cycle many of them become fiercely territorial and this means that, although their long-term goal is mating and breeding, they start out the season being rather belligerent. This is important when dealing with rivals in disputes over breeding space, but it can interfere with sexual encounters. It has to be suppressed to permit full sexuality to dominate.

In addition, prospective partners may feel afraid of coming close to one another, despite the sexual signals being offered. Most of the time animals prefer to keep their distance from other members of their species, even if their particular distance is only a few inches or feet. The final body-to-body contact that accompanies copulation is frightening, and this means that fear must be suppressed as well for mating to proceed.

So, in the courtship context, there are three conflicting moods: Fear, Aggression and Mating. This has been called the FAM complex and it takes different forms in different species. When aggression is strong and fear is weak, there are attacks between the male and female which become weaker and weaker as the sexual mood becomes aroused. When fear is stronger, one or other partner makes repeated attempts to flee, again with these becoming weaker as time passes and the mating act comes closer.

The dances and rituals of different species seem to have become fixed during evolution, according to the dominant moods of the species concerned. Fleeing movements have become stylised, or attacks have become modified, so that now they present only 'operatic' versions of the full behaviours. Frequently the movements are performed with a rhythmic pattern or repetitions and become stereotyped in the way they are executed, and these patterns vary quite distinctly from species to species. In this way they act as further safeguards against matings occurring between different species.

Two examples will clarify this process. Sticklebacks set up territories in streams and rivers in the spring. Each male fish builds a nest and waits for females ripe with eggs to come near. As soon as one arrives, she is attacked by the male, who is by now so worked up by fights with neighbouring males that he attacks anything and everything that dares to swim on to his small defended patch of riverbed. Males, or females without eggs, either fight or flee when attacked in this way, but the ripe female simply holds her ground and shows off her swollen belly. This excites the male and he darts away from the female and towards the nest where he wants her to lay her eggs. She begins to follow, but as she does so he dashes back towards her as if to attack again. Now he starts to dart, first towards the nest and then towards her, with each movement fixed and rhythmic in its execution. The male clearly has conflicting urges, to attack the intruder and to guide the female to the nest. Since the intruder and the female are one and the same, he cannot resolve his conflict. The outcome has been the evolution of the zigzag dance ritual, which successfully leads the female to the nest. Her presence there excites the male so much sexually that his aggression evaporates and he moves on to the next phase of his courtship which consists of showing her the nest entrance, nosing into it with his snout. If she responds by entering he becomes even more sexually excited. All aggression is gone now and he quivers ecstatically on her tail until she spawns, when he pushes through the nest after her and fertilises the eggs she has deposited there. The female now has a new, slimline figure. For the male stickleback this has no sex appeal whatever and he attacks her and drives her off his territory. He alone will now look after the eggs and rear the young.

This is a case of sex overcoming aggression and the stylised ritual of this process can be described as an fAM courtship. The zebra finch, mentioned earlier, demonstrates an alternative type, the FaM courtship, where the male's dance is a stylised version of his conflict between fear and mating. As he advances down the branch to his intended mate, he twists his body from side to side. With each pivoting movement, his tail twists round even farther than his body, as though he is trying to take off away from her while at the same time coming close to her. Each swing of his body brings him first closer, obeying his urge to mate, and then away from her, obeying his urge to flee. But as soon as he has twisted away from her he feels the urge to approach again. This pattern continues until he comes close enough to mount her and mate. The pivot dance display is now balletic and rhythmic – jump-and-turn-left/jump-and-turn-right – but its origins are clear enough. If it excites the female, she invites the male to mount by quivering her tail, and copulation takes place.

Nearly all the swinging, leaping, twisting and bobbing movements of dancing animals can be explained in terms of the urge to mate coming into conflict with fear or aggression. These conflicts, now evolved into rhythmically repeated sequences, have become vehicles to show off the bright sexual colours and have acquired an important new sexual role. A crucial element of many of the displays is that they are prolonged and repeated, and this gives the partners vital time together in close proximity during which they become not only aroused sexually but also perfectly synchronised in that arousal. When the moment of copulation comes, they are then both ready for the mutual sexual climax. Courtship displays may look comical to ignorant human eyes, but to the species concerned they are a crucial communication device at a key stage in the reproductive cycle.

When the wandering albatross, the world's greatest glider, that can easily circle the globe, comes to rest in the breeding season, it performs a series of prolonged dancing movements accompanied by strange groaning sounds. There is also a great deal of bowing and beak-rattling and, in particular, a head-tilting display in which the bill is pointed straight up to the sky. This appears to be an intention movement indicating that the bird is still slightly uneasy, which is not surprising considering that it has become unaccustomed to close company on its 4,000-mile journey to the breeding grounds.

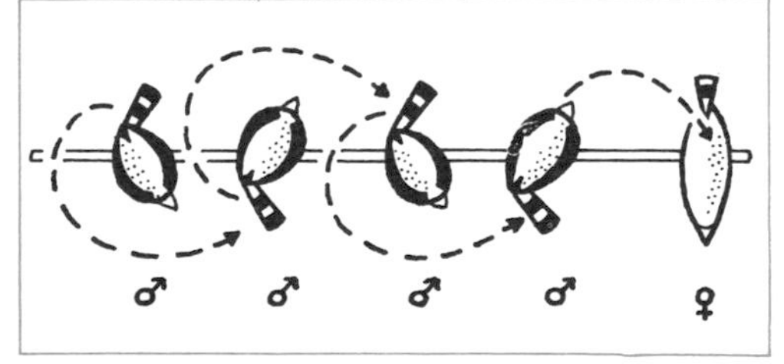

Some courtship dances are the result of a conflict between fear and sexual attraction. This applies in the case of the zebra finch.

The courtship display of grebes is another example of male and female synchronising their sexual arousal. Both members of the pair are fearful and aggressive as well as sexually interested, and the non-sexual feelings have to be eliminated by the courtship procedures. These involve a whole series of displays including synchronised running on the water (left).

Arena Displays

THE VAST MAJORITY of the 8,700 species of birds alive today form pair-bonds, with the male and female sharing the duties of rearing the young. Typically, the male establishes a territory, the female visits it and pair formation takes place. The male courts the female, a nest is built and the eggs are laid. Incubation duties and the feeding and protection of the young are shared by both parents.

This monogamous system works well for birds. Their parental task is a heavy one. By helping one another to get through the breeding season the male and female greatly increase the survival chances of their offspring. However, this is not the only system. There is a rare but dramatic alternative. In this, the male and female meet only briefly. They mate and then go their separate ways. There is a major division of labour, with the males spending most of their time displaying and the females carrying out all the nesting and parental duties. Only eighty-five species – less than 1 per cent of all birds – have taken this route, but they include the most bizarre and extraordinary avian species.

The reason why these non-pairing birds are so peculiar in appearance and behaviour is that the males, released from paternal duties, have been able to become sexual specialists. Their whole lifestyle has become consecrated to outlandish display, with the strangest plumage and the most unlikely actions to be found anywhere in the world of birds. As there is no pairing-off, each male can in theory mate with as many females as he can persuade to come near him. He is in direct competition with other males and it follows that, whatever the sexual signals are in his particular species, the more intense he can be, the better is his chance of gaining the female's attention. The male with the brightest feathers, the most explosive vocalisations, the most compelling courtship dances, will win the day. In this non-pairing world, the race is on for the super-male. All that holds him back is the danger of becoming too vulnerable to predation.

A top priority for a male of this type is to own a choice display ground, a specially selected stage on which to strut his colours. Females are attracted to places where there is more than one male to choose from, with the result that clans of these males tend to cluster together at traditional sites. These have been called arenas or leks (from the Scandinavian word *leka* meaning 'to play'). The definition of an arena is a communal mating station devoted purely to sexual activities, where no feeding or nesting

Ruffs (below, right) gather every year on traditional arenas where each male establishes himself on a small court. The area is visited by females who select a particular male for mating. She then leaves to nest and rear the young on her own. Here one (ruffless) female is seen surrounded by excited, displaying males.

Sage grouse (below) set up display arenas each year, where up to four hundred males gather. They display by inflating air sacs and calling with a loud booming sound. Females visit the arena and choose the best males for mating. A few top males – the best boomers – secure most of the matings. Direct, immediate competition between the males has led to an intense and extreme form of display.

occurs. The arena displays take place on personally owned courts or stages within the communal space.

One of the best known arena birds is the ruff, a European sandpiper in which the male in breeding plumage carries conspicuous head-tufts and a large collar of coloured feathers. In the spring, groups of these males gather on small hillocks in open meadowland. Each hill becomes the display arena for its group and individual birds establish themselves on small 'residences' that they quickly turn into bare patches on the ground. Each patch is only a foot or so across and there is usually a distance of about 3 or 4 feet between patches. For birds of this type, this is the full extent of their personal territory and they spend most of their time during

Arena-displaying males have all adopted a 'transformation display' technique as a way of impressing females. Each male suddenly enlarges himself in some dramatic way. The element of startling change is important because, like market salesmen, the males need to 'shout their wares'. Also, the species need to be quite distinct from each other, so that there is no confusion. The result is a set of amazing display colours, postures and structures, demonstrated here by the prairie chicken (top, left), the black grouse (top, right), the Kori bustard (bottom, left) and the blue grouse (bottom, right).

the breeding season sitting within it waiting for females to pass by.

At the start of the season there is much fighting between the males as they work out who gets the best residences. Females always seem to prefer the males near the centre of the arena – the top of the hill – and this is where the most dominant males finally end up and create their little bare patches. As soon as a female arrives, all the males adopt a stiff invitation posture with the display plumage fully raised. Looking rather like a bed of flowers they wait as the female wanders around from one to another. Finally she makes her choice and squats. The male mounts her, they mate and she leaves. The entire relationship lasts only a few minutes.

An oddity of the ruff arena is the presence of two types of male. They have been called *residence males* and *satellite males*. The residence males are the patch-owners and they have dark head-tufts. The satellite males own nothing and they have pale head-tufts. The satellites attach themselves to residence males and stay close to them. They never get involved in fights and are tolerated by the patch-owners. Why are they tolerated and what do they get out of their homeless state? Why should they exist at all?

The answer is that their presence helps to attract females to the residence near which they have positioned themselves. Females like to go where the males are thick on the ground – presumably to give themselves a better choice – and so it is to the advantage of the residence males to put up with the hangers-on. The advantage for the non-aggressive satellite males is that when the patch-owners dash off to attack rivals – which happens quite frequently – they can steal a quick mating with the female who is waiting patiently for their return. In this way, the unusual two-tiered male system is maintained.

The male of the Jackson's widowbird builds a special stage on which he performs. He fashions his stage by beating down the vegetation in a ring around a central clump. When the drab female arrives near by he flings himself into a frenzy of dancing and leaping while she quietly watches. After mating, she departs to lay her eggs, cuckoo-like, in the nests of other birds, which then rear her young.

Other open-country arena birds are the black grouse, the prairie chicken and the sage grouse. The males of these species all have elaborate displays with exaggerated feather erections, explosive vocalisations and dramatic air sac inflations. The sage grouse is the most extraordinary, with his pair of frontal air sacs bobbing up and down like huge breasts, and with his booming calls echoing across the sage-brush plains of North America. This species has large arenas up to half a mile long and 200 yards wide, housing as many as four hundred males, each on his personal patch of territory, with about 30 feet between each patch. As with the ruffs, the females seem to favour the centre of the arena, and nearly all the matings take place there, down the mid-line of the long, narrow arena space. The competition is furious among the sage grouse males, and only the very best boomers can hope to win the females, hence the amazing intensity of their displays. Of the four hundred males on one arena that was carefully observed, four master cocks managed to corner nearly all the females – 74 per cent to be precise. Nearly all the other cocks boomed in vain.

Arena displays are not confined to the open grasslands. They can also be found deep in the forest. The extraordinary, orange-coloured, cock-of-the-rock from tropical South America has male clans that share a general display area. This is divided up into many small courts – bare patches on the forest floor – owned and defended by individual members of the clan, who keep them clear of debris by violent thrashing actions of their wings. When a female approaches the arena, the birds fall to their patches in a swooping descent of breathtaking speed and proceed to flatten themselves there with their heads tilted sideways, showing off their 'Roman' crests against the dark earth. They continue to posture intensely for as long as the female appears interested and eventually one of them may be selected for mating. Although they fight one another to establish their personal patches on the display ground, whenever a strange male intrudes on to their general arena they gang up and attack him in force until he retreats. This indicates that despite competitiveness among themselves, they do also exist as a small 'display tribe', quite distinct from other such tribes.

The lyrebird males each hold separate territories inside which they construct about ten raised stages on the forest floor. They visit these every day, displaying with song and dance, their great lyre of feathers being laid horizontally over their backs, almost obscuring the birds beneath.

The male stagemaker, or tooth-billed bowerbird, collects large leaves and places them on his circular stage on the forest floor. The leaves are turned upside-down, making them much more conspicuous against the dark earth. The bird picks up and holds these leaves when he is displaying to his females.

Down in the forests of Australia there exists the equally remarkable lyrebird. The male of this species has amazingly long tail feathers and wire-like plumes. This bird differs in that each male has a larger territory within which he builds about ten small, circular mounds of soft earth on the forest floor, each about 3 feet in diameter. He visits these at regular intervals and, once on stage, begins his display, raising his plumes over his head like a decorative parasol, dancing and prancing and singing his loud song.

Also in the southern forests there are the exquisitely feathered birds of paradise. Some of these display on special branches, others on carefully prepared courts on the floor. In both instances a great deal of clearance goes on. The male magnificent bird of paradise, for example, spends most of his time trimming away the twigs and branches above his display patch until a bright shaft of sunlight can penetrate the dark forest, like a spotlight illuminating his stage. He can then perform his displays with his beautiful plumage shimmering with dazzling highlights – catching the eye of any female who happens to be passing near by.

Although these forest birds do not bunch together on communal, crowded arenas, they do still display on what have been called 'exploded arenas'. For, despite the greater separation of their personal territories, their loud songs are clearly audible to neighbouring males and, furthermore, the display courts are usually concentrated together in special areas of the forest. These become traditional display regions, to which the birds return year after year.

Perhaps the strangest of all the stage performers in the forest are the Australasian bowerbirds – eighteen species of the oddest animals on earth. The simplest of them is the tooth-billed bowerbird, or stagemaker. Like the birds of paradise, the male clears a patch of the forest floor. His circular stage is about 8 feet across and it is immaculate. He pecks and tugs at every tiny stem, root, leaf and twig until the whole area looks as though it has

The avenue builders, such as the fawn-breasted bowerbird (above), make collections of twigs and form them into parallel walls on a special clearing on the forest floor. In front of these bowers they place a collection of coloured objects on a flattened stage. They hold their coloured baubles in their bills when they display to their females, who often lurk inside the passageways of the bowers.

(Opposite) Each of the different species of avenue-building bowerbirds collects items of a specific colour when assembling piles of bower decorations. In addition each places them in a slightly different position. Lauterbach's bowerbird (top, left) collects red and pale grey objects, placing them in separate areas. The fawn-breasted bowerbird (top, right) collects pale green berries and places them directly in front of the avenue and also on the inner walls. The great bowerbird (bottom, left) places a huge pile of white objects directly in front of the avenue and pale green objects on either side of the white ones. The satin bowerbird (bottom, right) places bright blue objects all around the bower but not inside it.

been brushed clean with a large broom. It now stands out as a dark round patch in the forest, but this is not conspicuous enough for a bowerbird, for these birds are *collectors*. The male of each species seeks out special articles that he places on his display ground and which he employs during his sexual displays.

In the case of the stagemaker, his unusual, serrated bill is used to cut off huge fresh leaves from certain trees, leaves that are often much bigger than the bird himself. He manages about ten a day, carrying each one to the cleared patch of earth and carefully placing it there with his bill. He then fastidiously upturns the leaf to expose its pale underside. If a leaf is accidentally disturbed and flipped over, the bird immediately turns it upside-down again. A large stage can hold up to a hundred leaves and as soon as any one of them starts to wither or shrivel up it is quickly discarded and replaced with a fresh one. The display of these pale leaves against the dark earth is dramatic and no female, moving about the forest, would be able to ignore it. But presumably she only gives herself to the male who has the biggest and best leaves arranged in the most vivid and exciting manner and who, when she appears, can produce the best song and dance routine for her, as he leaps about his stage, clutching first one and then another of the big leaves in his bill.

Even more remarkable are the eight species of avenue-builders. These bowerbirds are also avid collectors and males of each species have their own favoured colours for the objects they bring back to their bowers. The spotted bowerbird loves white things and the display stage of one individual was found to contain over a thousand small white bones, and also white pebbles and stones, and white snail-shells. The fawn-breasted bowerbird likes green objects and the satin bowerbird favours bright blue. It is this last species that has been studied most closely and we now know a great deal about the construction of the male's bower and the complex way in which he uses it.

The satin bowerbird is found in the eastern forests of Australia. The male is an intense, shimmering blue-black with bright blue eyes. The female is dull and camouflaged. To construct his bower, the male clears a patch of ground and covers it with a layer of coarse grasses and twigs. At one side of it he builds an avenue of vertical twigs – a passageway about 5 inches wide with a wall of twigs down either side. The parallel walls are about 12 inches high and 4 inches thick, made of carefully collected twigs which are wedged down into the flat platform beneath. They are arranged in such a way that they arch over slightly, giving the passage a half-covered look.

On the display stage in front of the avenue the male then places a motley collection of coloured objects. They are never put into the passageway itself but are confined to the flat display area at its northern entrance. The avenue is always arranged in a north—south direction, with the result that when the male displays on the stage, with the female watching him from inside the avenue, his feathers will catch the sun and he will shimmer all the more fetchingly in front of her.

When he is setting up his bower at the beginning of the breeding season, he travels far and wide in search of his favourite colour. He brings back blue parrot feathers, blue flowers, blue berries, fragments of blue glass, broken pieces of blue crockery, blue rags, blue buttons and even blue bus tickets, if he lives near to civilisation. His special prize is the blue-bag he can sometimes steal from a local laundry. In captivity, satin bowerbirds kept in mixed aviaries become so frustrated that they have been known to kill small blue birds living there and place their corpses carefully on the grass stage.

Every day he runs a careful check on his collected objects. If any have lost their colour during the night they are discarded, and if anything red

should happen to have drifted on to the hallowed ground it is rapidly scooped up and carried away deep into the forest.

At dawn, the male's performance begins. Watched by the female he positions himself on his display arena and carefully selects one of his coloured objects. Picking it up in his bill he starts to make a strange whirring noise. Then, with his blue eyes bulging, he fans his tail feathers and starts to flick his tail and his wings in short, sharp movements, his head held low with his neck stretched out. His movements seem threatening and, as he shifts the position of his body, his plumage shimmers and glistens in the dappled sunlight. The female is attracted by these actions and comes closer, often standing inside the passageway of the bower where she continues to watch him but takes no active part in the proceedings. Mostly she is silent and appears to be idling the time away, but occasionally she gurgles softly and sometimes gives a small start when his actions are particularly forceful. The curious feature of his actions is that they are not directed towards her. As he leaps and hops stiffly about his display arena, picking up first this and then that object in his bill, he seems to be ignoring her and focusing all his concentration entirely upon the coloured articles he has collected so painstakingly. And yet if the female suddenly departs, the male stops displaying immediately and starts calling to her until she returns. Clearly, she is important to the display, even if it is not performed towards her.

Although these actions are odd enough, the strangest is yet to come. When the male is not displaying in his arena, he spends a great deal of time repairing and improving the walls of his bower passageway, pushing and shoving the vertical twigs more firmly into the ground, and fiddling with their positions. The finishing touch is to *paint* the inside walls that line the passage. This interior decoration is achieved in an extraordinary fashion. The bird searches out burnt wood from which he takes a billful of charcoal. This he mixes with saliva inside his mouth. He then picks up a small piece of bark and works it into a soft, sponge-like wedge that he holds in his bill. The wedge keeps his mouth slightly open, allowing the saliva to be dribbled out of the gap as the bird wipes his 'brush' over the surfaces of the inner walls. This painting activity covers the walls of the bower with a thick black plaster. Unfortunately, heavy rain soon washes it away and the male must redecorate almost every day. Bush fires leave so many burnt logs in the forest, however, that he can usually find plenty of black pigment without too much difficulty.

Some birds employ blue fruit-juices instead of charcoal as the colouring material for their saliva-painting activities. Blueberries, plums and other fruit-pulps are used and, on one occasion, one of the stolen blue-bags was squeezed to produce a wonderfully vivid blue band of colour on the inner walls of the avenue. It is odd that the male, who will not tolerate a single blue *object* from his display stage being placed inside his bower passageway, nevertheless plasters it with blue paint. There is certainly nothing simple about the breeding behaviour of the satin bowerbird.

How can this complexity be explained? What does the bower mean to this bird? It is not a home, a roost, a nest, a bed, a defence or a feeding site. It is purely and exclusively concerned with sexual display. Why does the male go to the enormous trouble of building and decorating it, instead of simply singing and dancing to his female like so many other bird species? The female makes no use of the male's twigs when building her nest, so what is all his building activity prompted by?

The answer seems to be that the bowerbird's bower is an extension of his own body. The other forest stage-performers – the birds of paradise and the lyrebirds – have bodies literally extended, with greatly elongated plumes and ornate feathers, but this makes them cumbersome. The bower may be cumbersome, too, but its owner does not have to carry it around all

(Top) The male satin bowerbird (on the left, with the paler, blue-eyed female in the avenue) may not place objects inside the bower but he still decorates it. He does this in a most unusual way, by mixing coloured fruit juice or charcoal with his saliva and dribbling this 'paint' down a paintbrush stick held in his beak. He wipes this brush on the inside of the bower, painting its walls. Here the twigs are seen coated in a black, shiny paint (below).

The golden bowerbird builds himself a complicated structure on a pair of saplings.

the time. When he takes flight into the forest he is still a small, sleek, athletic bird, not a stage star in fancy dress. If the bower can excite the female, then it is perhaps a more efficient way of making a dramatic display. Its drawback is that it is immensely time-consuming, compared with growing and cleaning giant feathers. But even here there is a hidden advantage. For the male can use it to preoccupy the female and trap her attention long before she is ready to mate and build a nest. It seems to be a case of 'the early bird catches the female', with rival satin bowerbird males building and decorating their bowers earlier and earlier, until, during the course of evolution, they have become fully active weeks before the female is prepared to copulate. The characteristic of the male displays is that they startle the female and give her more and more striking memories of particular males in particular bowers, so that when the mating time eventually comes she is likely to return to one rather than to another.

The building of the avenue appears to be primarily a way of manoeuvring the female into the best position to watch the male sing in the sunlight. And his painting of the *inner* walls of the bower with charcoal and blue juice is a move that impresses his blue-black colours on her even more vividly. These are the walls right next to her eyes as she watches him display.

In origin, the painting activity can only be courtship-feeding or nestling feeding, just as the building of the bower must have developed from nest-building. Like all stage-performing birds, the satin bowerbird male takes no part in the real nest-building or the feeding of the young, but that does not mean that these activities have become completely dormant in him. His bringing of twigs, his collecting of small objects and his dribbling of saliva down the walls of the avenue are all reminiscent of the actions carried out by paternal birds of other species that collect small food objects to bring to their mates or their nestlings and help in the construction of the nest. The difference here is that no eggs are to be laid in this 'nest' and it can therefore be elaborated to suit its new role as a display stage for the male.

Because the bowers need never hold eggs, their shapes vary wildly. In addition to the avenue-builders there is a group known as the maypole builders or gardeners. Here the constructions are even more impressive – in the case of the golden bowerbird they can reach an astonishing 9 feet in height. Some of the gardener birds build what look like native huts, with central supports and roofs sweeping down to side entrances. In front there are display stages covered in moss and decorated with coloured flowers, fruits and berries. When these strange buildings were first seen, deep in the Australian forests, their discoverers refused to believe that they had been made by mere birds, insisting that they must be some kind of secret grottos of the Aborigines. We now know of course that they are all the work of the amazing bowerbirds.

Although these display activities are genetically controlled and rely little on individual learning or cultural traditions, they nevertheless present a remarkable parallel with certain aspects of human society, where we too have externalised so much of our behaviour, going beyond our bodies in our displays, using clothing and building as a way of expressing our status and our moods. It shows what can happen when an animal that only possesses a bird brain starts a similar competitive trend. And it gives us pause to wonder just how much of our own building and decorative behaviour is genetically rather than culturally controlled.

Mating Behaviour

Once the rituals of courtship have achieved their goal and brought together male and female in an excited sexual condition, there remains the difficult problem of ensuring physical contact between their eggs and their sperm. Because the synchronised arousal of courtship has made both partners so ready to mate, the consummatory act of fertilisation might seem easy enough to accomplish, but it is not. The reason for this is that the all-important sexual cells are so small and so vulnerable. Protected deep inside the bodies of the mating pair, their frailty puts special demands on the behaviour of their owners. These demands have been met by two contrasting strategies: the external and the internal.

External fertilisation is crude and wasteful, but has none the less proved successful for many lower forms of life. At its simplest, it consists of synchronised shedding of eggs and sperm. The males and females, reaching their sexual climax at precisely the same moment, shed their sexual products into water, where they mix together in a dense cloud. Thousands of fertilisations take place all at once, and the fertilised eggs must then face the hazards of a hostile hungry world without any special protection from their parents. The majority of these tiny, growing life forms will be eaten, but a few will survive – just enough to make this clumsy method viable. This is success by sheer strength of numbers and its only serious problem, for the mating adults, is the delicate timing of the operation. Once the eggs have been shed into the outside world, they must soon start to harden and form a protective outer layer. The sperm have to make contact quickly, before this process has begun. External fertilisation is therefore an urgent matter. The males must be close to the females at the very second when the mass of eggs is expelled.

Many fish employ this mass-production technique of external fertilisation, nearly always on special spawning grounds where the males and females gather at breeding time. The numbers of eggs involved can be huge. There is one North Sea site, for example, where fifty million plaice gather each winter to mate and where the total number of eggs involved may be as high as 8,750,000,000,000. As soon as they have been fertilised by the males this vast cloud of eggs disperses by floating to the surface and being carried away in the ocean currents. The many fish to adopt this free-floating method include herring, cod, eels, hake, barracudas, mackerel, tunny, gurnards and puffer-fish. Others, such as salmon and trout, have non-floating eggs that are laid in the gravel at the bottom of the rivers where the fish spawn after lengthy migrations. The up-river journeys are so long and so tiring – in some cases the fish cover 40 miles a day for two months to travel a total of 2,400 miles – that the adults are completely exhausted after mating and soon die.

The females of many of these gravel-breeders provide a little protection for their eggs by digging a shallow depression. They do this by beating at the substratum with flicks of their powerful tails, just before they are about to shed their eggs. Then the male and female swim close beside each other above the egg cavity and, almost as one, shed their eggs and sperm. The fertilised eggs fall into the dip and are quickly covered over by further actions of the female. After this, the developing young are left to fend for themselves with no further assistance from their parents. They remain in the protective gravel until they hatch, when they struggle to the surface and wriggle free.

Most species of amphibians employ a basically similar technique. Frogs and toads migrate to their ancestral breeding ponds, using a navigation method that we still do not fully understand, and, once there, form pairs. The males clasp the bodies of the females and remain on their backs until the mating phase is completed. This means that they are always present, in intimate physical contact, until the vital moment when the eggs are shed.

Many animals favour mass production of eggs and leave their offspring to fend for themselves. Sperm and eggs are shed externally, are quickly fertilised and then lie in the water unattended. Many will perish but because of the huge numbers enough will survive to create a new generation. Throughout the breeding season male frogs and toads, such as this brightly coloured golden toad from Costa Rica, cling tightly on to the back of their female, ready to shed their sperm on the eggs as soon as they appear from her body.

Newts use a variety of unusual mating postures. Here, the male pleurodele newt clasps his female from underneath her body. He does this by hooking his front legs up and around hers, leaving his back legs for swimming. When he is ready to mate, he deposits a sperm capsule on the substratum and then makes a sudden sideways movement that causes the female's cloacal region to drop down on top of it. The sperm are then taken up into the female's body, where fertilisation takes place internally.

The male responds immediately with an ejaculation of sperm and the masses of fertilised eggs are then deposited in the breeding ponds as great lumps or long strings of spawn.

Throughout the lengthy clasping phase of frog and toad mating, the male task of clinging tight to the female's often slippery body is eased by the growth of 'nuptial pads' – rough-surface patches of dark skin that develop on the front feet. Sometimes there may also be rough skin on the chests of the males, helping to prevent slipping, especially when the mounted male is attacked by rivals desperate for a female of their own. Certain species have females that are so immensely fat and spherical that the unfortunate males are simply not capable of clasping them with their front legs. For them an alternative mating device had to evolve. It took the form of a kind of glue – a highly adhesive substance secreted by glands on the belly of the male – which sticks him to his female so powerfully that they cannot be torn apart without seriously damaging their skins. Only after mating is over is it possible for them to become separated.

Instead of individual pairs laying their spawn at separate parts of the ponds and lakes where they mate, in many species eggs are carefully amassed in one vast platform of spawn that may stretch for many feet. The advantage of this concentration in one area is that it provides a protection against cold, especially for the eggs in the middle of the platform.

One step up from this crude shedding of loose eggs is the careful depositing of spawn on special surfaces. Sticky eggs of many fish become attached to the rock-face in small cavities, where they are safe from disturbances in the current. In this protected position, they can grow and flourish and they can, if necessary, be tended by one or both of the parent fish. The mating act in such species requires the careful selection of a

Salmon swim thousands of miles to spawn in their ancestral rivers. Once they have arrived at the shallow breeding grounds the females dig a slight depression in the gravel by turning on to one side and beating downwards with their powerful tails (above). They are capable of shifting quite large pebbles in this way and once the cavity has been formed the male and female lie side by side on top of it and shed their eggs and sperm simultaneously. Trout mate in a similar way, and the species seen spawning here is the brown trout (below).

suitable surface for egg-laying and then a sexual manoeuvre that brings the bodies of the male and female close together at the chosen spot. For cave-dwelling species like the river bullhead, this also means establishing a defended territory. The male selects a cavity beneath a stone, clears it out and enlarges it by digging with his mouth and then lies in the entrance and awaits a female. As soon as one passes near by he darts out and grabs her head in his large jaws. Turning, he spits her into his cave and then positions himself across the entrance, blocking her retreat. These caveman tactics have great appeal for the ripe female, who starts examining the roof of the cave, searching for the best spot to lay her eggs. Eventually, when both male and female are ready, they turn upside-down and lie side by side under the roof. As the female lays her spawn, the male fertilises it and the outer surfaces of the eggs then quickly harden. The female departs and the male remains to protect them from hungry egg-thieves.

Fish that live in a caveless environment must make other arrangements. Some species build nests of bubbles or plant filaments in which the eggs are laid. Where the nest is floating on the surface, the pair mate beneath it, curling their bodies intimately around one another. In this body embrace they become aroused to the point where eggs and sperm are shed and the fertilised eggs float upwards into the bubble-nest.

For some fish, the environment provides no possible site for a nest. If the eggs need protection, special measures are needed. The answer is to incubate the eggs either in a pouch on the parent's body or actually inside the parent's mouth. Both methods require a particular kind of mating act.

With seahorses, it is the male parent that carries the eggs. He does this in a pouch situated on the underside of his tail. Mating is performed belly-to-belly, with the female firing her two hundred eggs into the male's pouch as he covers them with his sperm. She uses an extended oviduct to do this and in some species has been seen to insert this into the male's pouch, making her appear strangely masculine in her mating role. The male keeps the fertilised eggs in his brood pocket for about a month, when the little seahorses are advanced enough to break out and swim free.

Several kinds of fish have adopted the remarkable strategy of rearing eggs inside the parental mouth. It is remarkable because it demands enormous restraint on the part of the mouth-brooding parent, who cannot feed during the incubation period. Despite the tastiness of fish-eggs and the increasingly sharp pangs of hunger, the devoted parent never swallows during the days when the eggs are developing and hatching. This technique

The river bullhead lays its eggs on the roof of its nest cavity, usually under a rock or stone. The male guards the cave and spits the female into it when she arrives. Once inside, she inspects the roof and deposits her eggs. While she is in the cave the male lies across the entrance, blocking her exit and driving away all rivals. After he has fertilised the egg clutch he must protect it for days, fanning it, cleaning it and defending it. (The roof of the nest cavity of the bullhead seen here has been raised by the photographer to show the egg clutch adhering to the rock surface.)

The males of some mouthbrooder cichlid fish have dummy egg markings on their anal fins. When the female sheds her eggs into the water she quickly snaps them up in her mouth where she will protect them and gargle them until they hatch. The male's markings convince her that she has missed a few eggs and she starts snapping at his fin, trying to scoop up the dummy eggs (left). Her contact triggers the male's ejaculation and she inevitably takes up his sperm into her mouth, where fertilisation takes place.

Gouramis construct floating nests by blowing hundreds of bubbles. The male and female mate directly beneath the raft of bubbles, wrapping their bodies around one another in a nuptial embrace. As they do this they simultaneously shed their eggs and sperm. The fertilised eggs can just be seen here (above) rising up towards the nest, where they will become embedded in the bubble mass and will be both protected from predators and well oxygenated. Another species that employs a similar mating strategy, the African pike, watches its floating raft carefully and guards it from attack (below).

of egg-protection demands an equally remarkable mating act. The male of the little mouth-brooding cichlid fish called *Haplochromis* fans a shallow depression in the sand, using deft beats of his fins. He then entices the female into this pit and courts her. When she lays her eggs she immediately picks them up in her mouth, snapping them up before he is able to fertilise them. There they would stay, infertile and useless, were it not for one extraordinary quality of the male. On his anal fin he carries markings that imitate the eggs – little orange spots that look so much like the real eggs that the female is convinced she has overlooked a few of the ones she has just laid. Trying to pick these up too, she starts biting at his fin, and this act triggers his ejaculation. He sheds his sperm and, as she continues to attempt to gobble up his 'dummy eggs', she automatically draws his sperm into her mouth. The actual fertilisation in these mouthbrooders therefore takes place orally.

Mating acts that involve the shedding of eggs and sperm are always risky. A sudden disturbance and all is lost. The answer is internal fertilisation – a mixing of eggs and sperm inside the female's body. To achieve this usually requires an act of copulation in which the genitals of the male and female are brought into direct contact with one another for the passing of sperm. This occurs in all higher forms of life – reptiles, birds and mammals – and many invertebrates as well, particularly insects. But there are a few special cases where the transfer of sperm from the male to the female, although not involving crude shedding, does operate by an indirect route. This is true of creatures such as octopuses and squids. The males have one tentacle specially modified as a sperm-carrier. In some species it is used to pick up capsules of sperm – the spermatophores – from the male genitals and pass them to the female. In others, the tentacle develops a fold that enables it to act as a tube along which the sperm packets can travel, to be inserted into the female's body. In other words, one of the male octopus's arms acts like a penis.

A similar device is used by most male spiders. They take the sperm packet into a specially modified leg called the 'pedipalp' and use this to pump the sperm into the female orifice. Other penis-like organs have been evolved independently in a wide variety of animals. To solve the problem of safely transferring the sperm, different parts of the body have been enlisted. In some fish, such as the tooth-carps, anal fins are rolled up into

Animals with a fixed abode, such as these barnacles, face special difficulties at mating time. They are solved in this case by the possession of an enormously long penis that can reach out to make contact with a neighbour. There is no danger of passing sperm to the wrong sex because barnacles are hermaphrodites.

At mating time the male and female squid clasp one another in an intimate, many-armed embrace. In these dwarf squid the male clasps the underside of the female, so that both animals face in the same direction. One of his arms is specially modified and equipped to pass a sperm packet, or spermatophore, to the female, inserting it deep into her mantle cavity. In effect, this means that the male uses one of his arms as a long penis.

tubes through which the sperm can be ejected. In sharks, the male possesses modified pelvic ventral fins that form a pair of hollow 'claspers'. During mating he inserts these, either both at once or one after the other, into the female's orifice and injects his sperm through them.

Hardly any amphibian males have penises, or penis-like structures, but in a few rare cases some kind of sperm-delivery organ has evolved. These have so far only been recorded in ten of the three thousand species. In one, it protrudes from the rear of the male, looking like a short, stubby tail, and has given the species the name of tailed frog.

Internal fertilisation is accomplished in most birds without the aid of a penis. The typical avian mating act consists of the male mounting on the back of the female and then twisting his cloacal aperture down and round to meet that of the female. With the two orifices in contact the sperm are quickly passed across from male to female and the pair separate. For some birds, however, this was not efficient enough and in a few instances male birds have evolved true penises. The huge ratite birds – the ostriches, cassowaries and rheas – are so big that the sperm need all the help they can get to shorten their long journey, and penises assist in solving this problem. These male sex organs are normally hidden, but can be protruded by the use of special muscles at the climax of courtship.

Birds that mate in the water face another difficulty. Unless the joining of the two orifices were perfect, water could seep in and interfere with the act of insemination. Counteracting this, ducks, geese and swans have all evolved a penis. Again, this is normally hidden inside the male's body but is protruded during copulation. This fact makes it a little easier to understand the ancient legend of Leda and the Swan.

Among reptiles, the possession of a male penis is almost universal. Only the primitive tuatara appears to be without one. That survivor of an earlier epoch mates like a bird, cloaca to cloaca, but all others mate with full intromission. The male organs are, however, of two very different types. The penises of tortoises and crocodiles are of the usual, single design that is also found in all male mammals, but the penises of lizards and snakes are rather unusual. The males of these animals each possess two penises. These are normally carried inside the body and are only extruded just before one of them is pressed into the female. It used to be thought that both penises were inserted, side by side, at the same time, and as a result they were called 'hemi-penises' or half-penises. But in reality only one is inserted and each is a complete penis but of a strange construction. Instead of being erected, like a mammalian penis, the lizard or snake penis is inflated by being turned inside out. It is brightly coloured, usually red or violet, and has a small groove down which the sperm pass. Over its surface are many strange details – hooks, spines, pleats, ridges and folds – which differ from species to species. Once the penis has been introduced into the female reptile's cloaca, it cannot be removed until the male has completed his sperm delivery, because the spines that emerge as it turns inside out lock it firmly into place. Even if the female decides to abandon the mating and starts to move away, she cannot separate herself. She merely drags the locked male with her. Whether she likes it or not, once she has allowed the male to mate with her she will be forced to experience a prolonged coupling. The time it takes the male to ejaculate is remarkable. It is often more than an hour and frequently several hours. The record is held by a rattlesnake that was observed to remain coupled for an astonishing $22\frac{3}{4}$ hours. This is longer than the mating time of even the most amorous of mammals.

Some snakes store sperm and use it at staggered intervals – one individual of the night adder species was observed to lay four fertile clutches of eggs at monthly intervals, following a single copulation. Delayed fertilisation of this kind is known from a number of different kinds of animals, including certain bats where mating takes place in the autumn but fertilisation does not occur until the following spring. The record again goes to a snake, however – an American cat-eyed snake produced a fertile clutch of eggs after an incredible delay of six years.

Mammalian mating follows a basic pattern in which the male mounts the female, makes pelvic thrusts, ejaculates and dismounts, but there is a great deal of variation in the time taken to do this. Australian marsupial mice have been observed to copulate for eleven hours continuously in one species and twelve hours in another. Some of the carnivorous marsupials are reputed to take even longer. At the other end of the scale, wild baboons in Africa spend no more than eight to twenty seconds mating, with an average of only six pelvic thrusts per mounting. Between these two extremes, small carnivores such as ferrets frequently spend as long as three hours coupled together, and even species as heavy as rhinos may remain mounted for up to half an hour.

The typical mammalian copulation, however, is a brief affair and the reason is not hard to find. When joined together the pair are vulnerable to attack and, for most species, it pays to mate quickly and then be ready for sudden flight if necessary. With intense sexual arousal during the mating season this leaves the pair ready to mate again rather soon, and the usual pattern is for male and female to engage in a long series of many mountings during the time when the female is on heat. Wild rats have been seen to mate up to 400 times in ten hours and a small mouse called Shaw's jird copulated 224 times in a space of only two hours. A golden hamster was recorded copulating 175 times in one period of heat, each mating taking only a few seconds. Larger animals have longer periods on heat, and one female baboon was seen to mate 93 times with three different males during her oestrus phase that lasted for five days.

Wolves and dogs have an unusual mating pattern. The mounting is brief, but the male becomes 'tied' to the female so that when he dismounts he cannot separate from her. The pair must then remain standing together for up to half an hour, the male unable to withdraw his penis. Many observers have puzzled over the significance of this 'tie', which gives the impression of being both uncomfortable and inconvenient. While the pair are locked together in this way, they cannot flee from predators or from the attentions of other members of their social group. Occasionally the female may try to pull away, but simply ends up dragging the male with her. The reason for the tie is that the male in this family of carnivores takes many minutes to complete his ejaculation. Ensuring that this is done, his penis swells up once it has been inserted and does not deflate again until the ejaculation is over. During its swollen phase it cannot be withdrawn, no matter how hard the couple struggle.

In cats there is a different system. There, the females do not ovulate until they have been mated. The act of copulation triggers the ovulation and makes certain that fertilisation will take place. But a mild mating is not enough. In order for the female to respond, the mating act must be intense and, indeed, painful. The shock of it is then enough to start the process ovulation. This is achieved by the presence, on the tip of the male's penis, of some sharp barbs that point backwards. They do not interfere with the insertion of the penis, but they do make it difficult to withdraw. When the male does dismount and pull himself away, the spikes on his penis tear the inside walls of the female's genital passage and this makes her cry out and twist round to attack the male. Experienced males become adept at anticipating the swipe of their partner's paw and leap back as quickly as they can. However, far from turning the female against sexual encounters, this sharp pain of withdrawal has no lasting effect on her mood. In a short while she is again ready to accept the same male or another one for a further mounting.

A rare example of an amphibian with a penis-like structure is the tailed frog from North America. When mating, the male (underneath the female in this photograph) curls his short 'tail' between the legs of his mate and in this way manages to achieve internal fertilisation. This is important for this particular species which inhabits fast-flowing streams where sperm-shedding would be useless.

The mating behaviour of spiders is unique. The male employs his modified front legs to transfer his sperm to the female. These legs, called pedipalps, have tips that are specially adapted to take up a sperm packet from the male's genital opening and, following an elaborate courtship, pass it to the genital opening of the female. To achieve this manoeuvre, this wolf spider has mounted his large female backwards and is in the act of reaching down to the underside of her abdomen with his sperm-laden pedipalps.

Internal fertilisation is much safer than the external shedding of eggs and sperm, but it creates its own special problem – how to introduce the sperm into the body of the female. The design of many animals is not ideally suited to this task and copulation is not always an elegant or easy matter. For hard-shelled turtles, the clumsiness of the operation sometimes leads to unwanted competition (above). For monkeys, such as these crab-eating macaques (opposite, top), there is the difficulty of masculine legs that are slightly too short for the job. Many species, so well adapted to other pursuits, such as running, leaping, swimming or flying, find the moment of copulation difficult to achieve with any degree of grace or deftness.

The mating behaviour of lions and other members of the cat family is a painful process for the female. The male's penis is barbed and its withdrawal hurts the female, who frequently twists round to attack the dismounting male (opposite, bottom). The pain is necessary for feline mating because it is this shock to the female's system that induces her ovulation and permits fertilisation.

Many male mammals can be seen to grab their mates by the scruff of the neck when mating. This neck-biting looks aggressive and even brutal, but it is not. In fact, it is a small behavioural trick played by the males on their females, enabling them to mount without too much difficulty. The neck-bite acts as a pacifier and immobiliser. To find the reason for this, it is only necessary to look back at what happens to young mammals when they are being transported by their mother. She picks them up by the scruff of their necks, holding them firmly but gently in her jaws. Their reaction to this is to lie immobile and to allow themsleves to be transported without struggling. This is important if they have to be quickly removed to a safer place. When the adult male mammal grabs his female in his jaws, she reacts as if she were his offspring. Programmed in infancy to lie still and not struggle when her mother bites in this manner, she automatically responds in the same way to her mate. In this fashion infantile responses can be usefully exploited as a technique that ensures a smooth copulation.

In all the examples mentioned so far, it has been taken for granted that a male is a male and a female is a female and shall always remain so. This is true of most forms of animal life, but there are intriguing exceptions. Many worms and snails are hermaphrodites, each sexual partner carrying both male and female sex organs. When they mate it is, so to speak, a double marriage. With some slugs and snails there is an elaborate courtship preceding the mating act, and this creates the problem of who 'leads', since both partners are male and both are female. The answer is that one partner acts the male role and one the female during the preliminaries. But they often swap roles from one mating to the next. As a refinement of their mating act, each partner carries a sexual weapon in a small internal sac. This takes the form of a small, sharp dart, made of the

Many lower forms of life are hermaphrodites, so that the mating act is a double copulation. This is true of common earthworms (below), snails (right), and slugs (bottom). The mating behaviour of slugs is spectacular. The pair begin by circling one another, high up on a branch or top of a wall. The circling becomes tighter until they are caressing one another with their tentacles. This may go on for as long as ninety minutes, with both animals secreting large quantities of mucus as they become increasingly excited. Then, when they have formed a strong mucus string from their circling, they entwine their bodies and lower themselves, hanging from their mucus lifeline. This hanging thread may be up to 18 inches in length. Now they extrude their long, pale penis-sacs, which hang down below their heads. As they become more aroused, these sacs change in appearance from being club-shaped to being fan-shaped. At this point the sacs themselves become intertwined and change shape again, developing lobes, and the climax of sperm exchange takes place.

same substance as ordinary snail-shell, and during courtship the animals fire their darts into one another's bodies. In some species the darts are large and painful; in others they are smaller and more mild in their impact. For each species the pain has to be at just the right level for it to be sexually stimulating. In this way there can be no error regarding the identity of the sexual companion.

Another mating system employs separate males and females in the more usual way, but has each individual changing sex at some point. In certain shrimps each animal starts out life as a male and then changes into a female. In certain oysters the process repeats itself many times, with one oyster switching from male to female several times in a single year. Such changes are rare in higher forms of life, but can be found in creatures as advanced as fish. The Mexican swordtail is the best known example. There the males are smaller than the females and they arrange themselves into a masculine hierarchy or peck-order. Usually, the biggest males are at the top of the social order, so it follows that if a female can change sex she will automatically rise to the pinnacle of social power by virtue of her much heavier body. So in any community of swordtails it is possible to see the largest of the little male-males being dominated by the even bigger female-males that have already produced young and have now switched to the masculine gender for the second half of their lives. The intriguing question that remains to be answered is why do some species show regular gender-switching or hermaphrodite matings, when the vast majority of animals indulge in more straightforward and simple male – female matings throughout their lives?

Nesting Behaviour

A YOUNG GIRAFFE is 6 feet tall when it is born. Its mother stands still when giving birth and the long-legged baby crashes to the ground in a heap. In a few seconds it is struggling unsteadily to its feet and tottering around near its mother's body. In no time at all it is following her as she moves off and its life on the open plains of Africa has begun. For such animals there is no need of a nest in which to rear their offspring. Birth is delayed until the young are ready to run and protect themselves from danger. But for many other species the first days of life are much more vulnerable and parents must provide a snug home for the young if they are to survive. Nests come in many forms, some of which are marvels of engineering and architecture.

The classic nest – the one that immediately springs to mind – is the cup-shaped structure favoured by the majority of bird species. Fashioned as a hollow hemisphere it is ideally suited to contain eggs, with the parent bird sitting on the clutch. It has the disadvantage of being vulnerable to predators, but the advantage that it gives the incubating parent bird a clear view of the surrounding environment. The approach of an enemy can easily be seen, enabling the bird to flee swiftly at the last moment if necessary.

One step up from this is the domed nest. Spherical in shape and with a side-entrance, this provides greater protection from the sun, rain and cold and it completely hides the parent from prying eyes. But it also makes it impossible for the parent to know when a predator, such as a tree snake, is closing in. Last-minute escape is therefore more difficult.

A further advance in design is the tube-nest, the entrance to the domed nest being extended downwards as a long, hanging passage. With the nest attached to a slender twig or branch, this design makes it virtually impossible for tree-snakes or other small, climbing predators to enter the nest-cavity. The problem with this type of construction is that it requires a great deal of time and effort to fashion it. The stitching of the nest-fabric involves deft beak-and-claw actions of a highly specialised kind. These have been perfected in the group of small seed-eaters called weaver-birds.

The male weavers begin their painstaking task by flying to tall grasses and tearing off long strips. They do this by nipping a blade of grass and then flying sharply and powerfully upwards, holding the blade firmly in their beaks. Making sure to take the toughest strips first, they start to wrap them around the fork of a twig. After a number of strips have been collected, the loose ends hang down in two separate trailing tassels. These are then brought together to form a circle and this becomes the architectural basis for the whole nest. More and more strips are brought to this ring, the male standing on it like a caged bird on a swing as he carries out his intricate weaving. Stretching out as far as possible with each building action, the bird gradually forms the dome of the nest. He uses three main types of stitching: knotting, weaving and twining. When knotting, he holds the grass with one or both feet and then pulls, pushes and twists it around with his beak. He is capable of tying slip knots, overhand knots and half-hitches. When weaving, he performs the complex movements of a craftsman making a wicker-basket as he forms and improves the fabric of his remarkable nest. When twining, he uses his beak like the fingers of a tailor sewing, threading a new grass strand in and out of the nest-wall and adding strength by employing a variety of loops.

All weaver-bird nesting is carried out by the more colourful males, the dowdy females taking no part except to bring a few soft grasses for use as an internal lining. Each male displays at the entrance of his nest. The females visit the nests and finally select one for egg-laying. After they have mated, the pair separate, the female remaining inside the nest to rear the young while the male moves on to construct another nest and perhaps attract a second mate. If he fails to procure another partner, he destroys

The cup-shape is the classic design for a bird's nest. It has the disadvantage of exposing the parent bird – in this case, the bullfinch – to the elements, but the advantage that the adult can scan the horizon for danger while it incubates the eggs.

Weaver birds construct domed nests that hang from slender twigs. The basis of the nest design is the initial ring (above, left), around which the whole of the structure is formed (above).

The tailor-birds of Asia (opposite, top) construct inaccessible nests by stitching together large leaves and filling the created space with fibre and soft down.

Social weavers build communal nests (opposite, bottom) that are larger than the nests of any other bird species. Each 'apartment block' contains up to a hundred separate entrances to individual nest-holes, with all the entrances facing downwards. Sometimes these nests become so huge that they collapse under their own weight, especially after they become sodden in a heavy downpour of rain.

his nest and starts again, often building many separate nests during the course of a season.

Some weaver-birds cooperate to construct huge communal nests. Instead of making individually woven units, they combine their energies and work rather like human roof-thatchers. The result is a massive arboreal haystack built around a forking branch. Inside its bulk each bird makes his own separate cavity with a downward-facing entrance-tunnel, and there can be up to a hundred such units in a single communal nest. These are the largest constructions made by any bird species, some of them measuring as much as 20 feet by 13 feet by 3 feet. They provide great security for the social weavers that assemble them, but they do have one crucial weakness: after heavy rains they may become so sodden that they collapse under their own weight and crash to the ground.

There are many other remarkable nests to be found in the world of birds. In the New World the orioles build hanging nests similar to those of the weaver-birds, but they add a special refinement: they locate them near wasp nests where monkeys will be reluctant to raid them. In desert regions the tiny elf owl often nests in holes in giant cacti, a site that is secure from almost any enemy. In Asia the little tailor-birds stitch together the sides of a large leaf, using knotted fibres that are passed through special holes made in the leaf-surface. Their small, snug nest is then formed inside the inaccessible manufactured container.

Many birds use some kind of 'cement' to strengthen their nest structures. One of the Indian Ocean seabirds, the lesser noddy, employs only slippery seaweed to make its shallow nest but cements this to the supporting branch with its own excreta. The scissor-tailed swift has saliva so adhesive that it can be used to fashion a heavy hanging entrance tube beneath the nest-cavity. And the famous constructions of the edible-nest swiftlet, used in the making of bird's nest soup, are formed almost entirely

from the solidified saliva of that species. Those who choose to savour this kind of soup must face the fact that they are in reality dining on avian spittle.

Perhaps the most amazing nest of all is the one created by the Cape penduline tit. This small bird builds what looks like an ordinary domed nest with a large side-entrance, hanging from a slender branch. It is finely made and appears to be safe enough but its security system is even more complicated than outward appearances would suggest. Its special refinement can only be discovered by examing the interior of the nest. It can then be seen that the large side-entrance is a fake, leading to a false cavity. The real entrance is a narrow slit that lies just above it. This slit closes tight whenever the bird squeezes into or out of the real nest-cavity, a space that lies beneath the false one. This is a construction that would defy even the most intelligent of egg-thieves.

Some birds have almost given up nest-building and do little more than find a suitable notch on a branch or a tree-trunk, where the egg is laid and perilously incubated. The tree swift cements a little patch of material to a branch with its saliva, just enough to keep the single egg from rolling off. The parent is forced to cling tightly to the branch throughout the whole of the incubation period and the young bird must do the same during its entire nestling phase. The potoo, a relative of the nightjars, goes even further. It builds no nest at all, merely selecting a small notch in the trunk of a tree and depositing a single egg in it. The camouflaged parent incubates this egg in a stiffly vertical posture, looking very much like a small outcrop from the main trunk. The young bird, after it hatches, adopts a similar rigid pose.

If these birds have reduced nesting to the absolute minimum, the mallee fowl, or mound-building bird, of Australia has gone to the other extreme,

Some birds prefer to take over the old nests of other species. The little American elf owl (right) often nests in a giant cactus where a woodpecker has already fashioned a convenient cavity.

The ingenious nest of the Cape penduline tit, with its false entrance and its concealed true entrance, is perhaps the most remarkable of all avian nest constructions. It employs a design principle not unknown to the pharoahs of Ancient Egypt as they planned their secret burial chambers. These photographs (below) show the bird in front of the false entrance and the bird entering the inconspicuous, true entrance above.

its huge incubator being one of the most laborious and demanding of all avian parental devices. Instead of sitting on the eggs like other birds, this species builds a vast mound in which the eggs are buried. It is the male that carries out this gargantuan task, a parental duty that keeps it busy from dawn to dusk. Having started out by digging a deep pit, the bird then scratches towards it all the loose leaves and twigs from a 50-yard radius around it. These are packed tightly into the centre of the pit and covered with sand. The male has now, in effect, become the owner of a large compost heap. Inside it the decomposition of the vegetation creates a dramatic rise in temperature and when the female arrives she is encouraged to lay her egg deep in the centre of it. Each time an egg is deposited, the male examines the heat of the mound with extreme care by burying his head in the sand. After all the eggs have been planted, his task is far from finished. In fact, it is only just beginning, because he must now keep a constant check on the temperature of the mound. Whenever it rises too high, he must open up the mound by scratching away a certain amount of the sand. This lets out some of the decomposition heat. If the eggs become too cool, he must try to heat up the mound. He does this by spreading out some of the top sand, letting the sun warm it up, and then piling it back again, close to the eggs. He is constantly on temperature-duty until the eggs hatch, when the tiny chicks have to push their way up through several feet of sand to reach the surface. Once there they must fend for themselves for, amazingly, the mallee fowl does nothing to protect them. After spending day after day labouring for their successful hatching, he totally ignores them once they are above ground. So too does their mother. The young flee from their parents and hide in the undergrowth if they happen to see them. If cornered by a predator they are ready to fight with both beak and claws when they are only four hours old. Since they learn nothing from their parents, it follows that the entire, complex business of mound-building, temperature-testing, and heat-controlling must all be completely inborn in this strange species. This underlines just how subtle and intricate behaviour-programming can be inside the brain of a bird.

The mallee fowl is not the only animal that builds an egg-incubator. Surprisingly one can also be found in the world of reptiles. No less a species than the mighty alligator follows a similar, if less elaborate, pattern. Here it is the female that constructs the compost nest, heaping up a mixture of mud and rotting vegetation in which she lays her eggs. While

they are developing inside it, she stands guard near by and defends it against hungry egg-thieves such as raccoons. When the young hatch out they must fight their way to the surface, like the little mallee fowl chicks, and, again like them, must fend for themselves once they have broken through to the outer world.

This is the only reptilian nest of any note, but a number of amphibians form mud or foam nests, often on a dramatically large scale. Masses of grey tree frogs cling together on branches high in the air, where they combine to whip up a foam nest from a secretion produced by the females. The eggs are laid in this foam, where they hatch into tadpoles and then drop down to the ground. Other species create huge foam nests at pond

The circular mound is a popular nest shape for many species, such as these flamingos (above), but the most dramatic mound-builder by far is the mallee fowl of Australia. The male of this remarkable species constructs a huge incubator mound containing compost. When the female arrives (top, left) she lays her egg in the middle of the mound, where the heat of decomposition helps to hatch it. The male opens the mound daily to check the temperature (top, right) and then makes adjustments to regulate it.

level, in which their young can grow and develop in comparative safety. Also at ground level, the Smith frogs build a circular mud nest that acts like a barricade, preventing the water inside it from spreading and drying out. They mate inside their circle of mud and their tadpoles develop in it, walled off from the rest of the world with its roaming killers.

Among the fish there are many cavity-nesters, but the most intriguing builders are the common sticklebacks. Each spring the males fashion a small nest from collected plant fragments and bore a tunnel through the centre of it. The nest-owner then leads a ripe female to the nest and encourages her to enter. Once inside she lays her eggs and departs. He follows her through, fertilising the eggs as he goes, and then starts his parental vigil. For day after day, without any help from the female, he tends, defends, repairs and maintains his little nest. His biggest problem is ensuring that the growing eggs have sufficient oxygen and he does this by spending hours in front of the nest-entrance each day, fanning a current of water through the tunnel with his pectoral fins. In the three-spined stickleback the nest is built on the floor of the stream. In the ten-spined species it is hung up in water-weeds. In both cases it is held in place by strings of glue secreted by the male as he shivers over and around it during its construction.

In regions where it is dangerous to leave spawn at ground level, certain frogs have taken to nesting in the trees. Some build leaf nests for their eggs, while others combine to create a huge foam nest (top) in which the tadpoles can develop in safety. When these tadpoles have grown sufficiently (above) they drop down to the ground and fend for themselves.

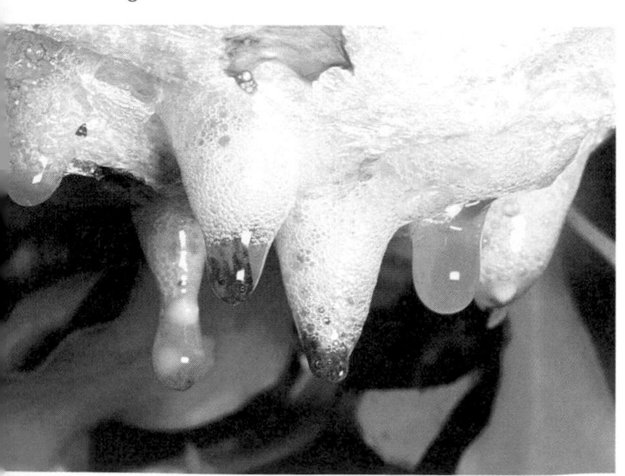

To find the largest of all animal nests it is necessary to turn, paradoxically, to one of the smallest of nest-builders: the termite. For these amazing animals the nest is their whole world, a dark, warm, humid subterranean kingdom in which their teeming millions labour incessantly to maintain an environment so complex that it makes the human space-colonies of science fiction look positively simple-minded.

There are two thousand species of termites, and many variations in the structure of their nests, but the most complex nests are constructed in the following way. During the rainy season, winged males and females break out of the old, established termite-hills and fly off in search of new homes. The vast majority are gobbled up by insect-eating enemies, but the few that do survive eventually shed their wings and pair up. The pairs separate off and go underground. There, each termite couple forms a nuptial chamber and mates. They care for their first offspring, but when they have created the nucleus of a new colony their young turn the tables and start to look after them. The royal couple continue to breed, with the female growing more and more in size, becoming a huge egg-factory and capable, at her peak, of shedding over thirty thousand eggs a day. Her faithful subjects carry these eggs away, through small tunnels, to brood chambers they have constructed near the central, royal cell. There, cared for and fed, they begin to grow. But, like all the other countless members of the colony, they never become sexually active. Only the original king and queen are capable of reproducing. All the rest – and there may be up to ten million of them in one colony – remain sexually stunted as 'pre-adolescents'. They develop into workers, soldiers or stand-ins. The workers and soldiers are almost blind and are concerned with the maintenance of the nest. The workers build the ever-growing nest structure, clean it, repair it and provide food for the colony members. The soldiers, with huge jaws, defend the nest against enemies. (It is this that chimpanzees exploit when they go termite fishing with twigs. The soldiers react to the twigs pushed into the nest by grasping them with their jaws, and the chimpanzees can then withdraw their 'fishing-rods' covered in clinging nest-defenders.) The stand-ins are individuals kept in a state of sexual readiness, should the original breeding couple die. When this happens a new pair is allowed to develop and the colony line can continue. Some colonies are known to have been in existence for over a century.

The architecture of the termite nest is extraordinary. A large nest can have a diameter of 100 feet and a height of 20 feet. Much of it is underground, but there are also impressive mounds protruding above the surface. These are manufactured from a mixture of excavated soil and saliva, and are incredibly hard and strong. The function of these termite hills is essentially that of air-conditioning. It is important for the interior of the nests to be maintained at a constant temperature and with the right degree of moisture. This is achieved by the building of a system of hollow cavities inside the hills that permit the dank, fetid air from below to rise and be dispersed through minute holes in the hard surface.

One kind of termite-hill, made by an Australian species called the compass termite, is a long, flat structure always built in a north – south plane. This is a special temperature-control refinement designed so that during the cool of the morning and evening the sun's rays fall on the broad side of the nest, helping to heat it up; but then, during the intense heat of the middle of the day, the sun shines not on the flat side but on the narrow end of the hill, thus preventing overheating.

In the depths of every termite colony there is a complex network of cavities and tunnels. Around the royal cell there are larval galleries and nursery chambers. Beyond these there are burrowing tunnels leading to food sources and above them are special fungus gardens, where the worker termites provide a substratum on which they can grow their fungal foodstuffs. Beyond these again are special food-storage chambers for the fragments of wood, leaves and plant materials that have been brought back by the foragers. Long tunnels may be built to reach far down to the water table below, as a way of maintaining the moisture level of the nest-cavities and also as a source of damp clay for wall-building. These tunnels may extend as deep as 100 feet if necessary.

In this way, the termites create for themselves a complex, highly organised and carefully regulated internal environment, teeming with activity and completely shut off from the outside world. Just occasionally, when conditions are right, a number of growing young are allowed to develop into a fully adult sexual condition and then the surface of the nest is broken open and they escape into the hostile upper world beyond, some to found new dynasties elsewhere. It has proved a highly successful nest-system throughout the tropics, failing only in the cooler regions to the north and south of the globe. The immense labour undertaken by these tiny insects can best be summed up by the observation that in one case a single nest was removed by human builders to manufacture no fewer than 450,000 bricks for use in human dwellings.

There is nothing on earth to compete with the termites for sheer industry – except perhaps our own species. By comparison, the nests of bees and wasps are rather modest, although in their detailed structure they are equally fascinating.

Honey bee nests contain, at most, only eighty thousand individuals, compared with the vast termite populations of millions in a single colony. The caste system is different, with one queen, a collection of several hundred drone males and the main body of thousands of workers. The workers have different duties, according to their age. Their first task is *cleaning* the nest; then they act as nurses to the young in their brood cells, *feeding* them initially with 'bee milk' from special glands, and then with pollen and honey. The 'bee milk', known as 'royal jelly', is only given to the growing young for three days if they are destined to become workers, but if they are to be new queens they are given a continued supply, a food boost that alters their development. At the end of their nursing period, the workers switch to *building*, manufacturing more and more of the hexagonal nest-cells from wax glands that become active on their abdomens. After this comes a period of *hoarding* in which the workers collect nectar and pollen being brought in to the nest and stack it away in food-storage cells. Next, the workers perform the important duty of

The sticklebacks build nests of plant threads and fragments. The three-spined species constructs its nest on the floor of the stream, weighting it down with sand (above) and sticking it together with a glue secretion. The ten-spined species builds a domed nest up in the water-weeds (below). All nesting is done by the males in these species.

The tallest nest in the world is built by one of the smallest nestbuilders. These huge termite nests can rise as high as 20 feet. During the rainy season winged forms develop in the termite colonies and break loose in a great swarm (above). The compass termite of Australia builds flat nests that are always aligned in a north-south direction. Seen from their flat side they look like tombstones standing in a deserted graveyard (above, right).

guarding the nest-entrance, always ready to sting any unwanted intruders. Finally, they move on to the last phase of their short lives, as *foraging* bees, out searching every day for pollen and for nectar to convert into honey. They may also occasionally collect resin, which is used to repair holes in the nest or alter the shape of the entrance.

The hexagonal cells of honey bees are assembled in vertical sheets, or combs, and these are constructed in groups inside hollow trees or other cavities where the wild bees can successfully defend their home against enemies. As with termites, it is important for the honey bees to keep the nest temperature constant. They prevent overheating by gathering at the entrance and fanning their wings. By standing facing into the entrance, their fanning action helps to draw the hot air out of the interior and encourages cooler air to seep in through small cracks and crevices. If this is not enough to reduce the temperature they go off in search of water. This they bring back and spread on the combs. They then fan the damp comb-surfaces, creating cooling by evaporation.

The intricacy and complexity of the nest-systems of the social insects is at first sight bewildering. It is hard to understand how they can possibly be organised by brains small enough to fit into the heads of termites, ants or bees, but we are becoming more and more knowledgeable about the control mechanisms that are used. Typically, these involve the timed switching on or switching off of the production of a particular chemical substance. To give one example, the queen bee produces a pheromone in her mandibular glands with the exotic name of *trans-9-keto-2-decenoic acid*. It is this single chemical that enables her to rule the roost, for it acts as a suppressant of sexual development in all her workers. If she failed to secrete it, chaos would ensue. When her powers start to wane, as she grows old, the strength of her chemical signal begins to weaken slightly and this triggers a new pattern of behaviour in the workers. They begin to build queen cells and to rear new young queens, one of which will eventually take over the nest. Also, if the colony becomes too large, the queen cannot suppress the whole community sufficiently, which has the same result. Queen cells are built and young queens hatch out, take off and start a new colony elsewhere. In this way a single chemical substance produced by a gland on a single occupant of a nest can influence the whole organisation of the social life of these communal insects.

Wild bees prefer to nest in a hollow tree, but if no cavity is available they will hang their structure of vertically arranged wax combs on a branch in the open.

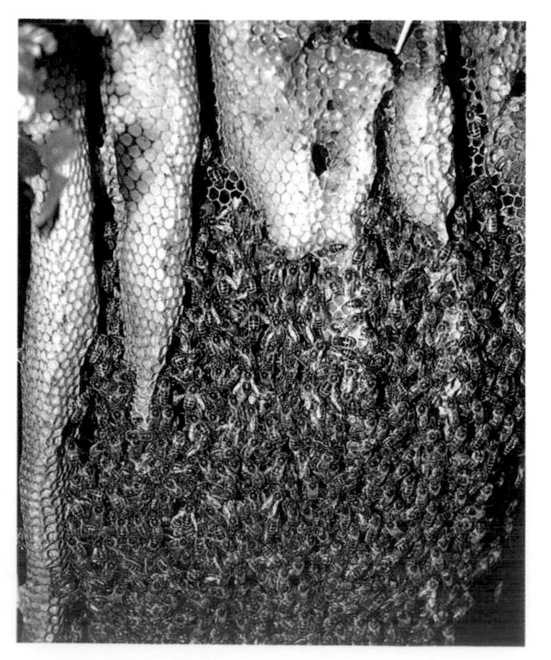

Parental Care

No species lavishes such prolonged or devoted parental care on their offspring as we do. In this respect, at least, the human species is second to none. But many other animals are also intensely active, both maternally and paternally, during their reproductive phases. They clean their young, they feed them, they transport them from place to place, they defend them against enemies, they prevent them from becoming too hot or too cold, they ensure that they have plenty of oxygen and they encourage them to learn about the world around them. Each species does not do all these things, but there are huge numbers that do at least some of them. Almost all birds and mammals, many fish and a few reptiles, amphibians and lower forms of life indulge in some form of parental protection and rearing.

One of the earliest acts of parental care is cleaning, especially with young mammals. As soon as they are born the mother nearly always licks them all over, examining them carefully and ensuring that they are completely free of the embryonic membranes. In many cases these membranes are eaten by the mother, as is the placenta, and the umbilical cord is bitten through and sometimes partially devoured. The mother frequently pays particular attention to the muzzle region of her offspring, an action that encourages the start of infantile respiration, if this has not already begun. The prolonged licking of the wet baby has the effect of drying out its fur and preventing chilling if the air is cold. Licking of the anal region is essential in many species if the young are to defecate and urinate successfully. The newborn animals are incapable of ridding themselves of body waste without this form of stimulation, and the mother often devours these products during the earliest days of life. In this manner she keeps the den clean and also removes odours that could attract predators.

An additional function of the cleaning of the new arrivals is that it provides the mothers with a strong memory of their individual identities. The personal fragrances of the young animals help mothers to distinguish their own offspring. Studies with sheep and goats have told us just how long it takes for a lamb or kid to become 'known' to its mother. In the case of sheep it requires twenty minutes of cleaning and nuzzling, but in goats no more than five to ten minutes. If the mother is only allowed to attend to her baby for a shorter period than this, she will happily accept a replacement for it. But if she has had her critical 'species time' she will no longer accept any but her own young. Later on she will add other clues to help her identify her growing offspring. In addition to its personal smell, she will recognise it by its calls and its appearance.

Many mammalian mothers continue to clean their young as they develop. Monkeys, in particular, spend many minutes every day searching through their youngsters' fur, removing dirt and particles of dried skin. These moments again help to reinforce the bond of attachment between parent and child.

For birds there is the special problem of keeping the nest clean. Nest sanitation is particularly important with nestlings that are eating large quantities of food. The parent birds are aided by a special form of defecation. The nestlings produce a convenient kind of dropping which is enveloped in a neat sac. This is removed from the anus of the young bird as it appears and the parent bird then flies away with it and drops it at a distance from the nest. Desert-dwelling birds do not need this form of parental protection against nest-fouling. Their droppings are so dry – to prevent wasteful water loss – that they can simply be stacked up at the rear of the nest.

The major problem for all attentive parents is providing enough food for the rapidly growing young. For birds this may mean over a hundred trips a day, to and from the nest. During this phase the nestlings appear to be little more than widely gaping mouths with stomachs attached. And as

The earliest parental activity is keeping the young clean. In mammals this means eating the embryonic membranes and licking the newly born offspring dry. This is done even by herbivorous species such as this springbok.

During rest periods, primate parents (above) spend a great deal of time examining the fur of their offspring, grooming them and studying their skin condition closely. This not only keeps the infants clean but also helps them to socialise and become more integrated with the rest of the group. Lionesses (above, right) are also attentive mothers, especially when boisterous play has made their cubs' fur dirty.

a way of ensuring that food is repeatedly crammed down their throats, the young of many species possess vivid mouth markings. Some have bright colours, others have an array of black spots and bars, while still others possess vivid swellings at the corners of their gapes. As soon as there is the slightest vibration at the nest, the scrawny necks are stiffly erected and the beaks gape as wide as possible. This signal is so powerful to the hard-working parent birds that they sometimes make a mistake. They become so responsive to gaping mouths that they may inadvertently stuff food down the wrong gullet. There is a famous instance of a small finch ramming a beakful of insects into the gaping mouth of an ornamental goldfish in a garden pond. The fish, accustomed to being fed by its owners, saw the bird's shadow and, sticking its mouth out of the water, received a meal from an unexpected source. The bird, seeing the gaping jaws, simply could not resist them.

Large seabirds help their nestlings digest their food by initiating the process for them. They catch a fish, swallow it, half-digest it and then regurgitate it as a warm fish soup. The young beg for it by pecking at the parental beak which, in some species, has special markings. The herring gull has a red spot on its bill which acts as the signal saying 'food is here'. If a herring gull chick is offered a red spot on the end of a thin piece of wire, it finds this highly simplified dummy more exciting than a realistic model of a gull lacking *only* the red spot.

Some birds, such as black vultures and others living in hot, dry regions, will carefully fly back to the nest with water in their mouths and dribble this down to their thirsty young. Pigeons provide a unique drink for their nestlings in the form of 'crop-milk'. The lining of the crop becomes so swollen with a white liquid that the cells burst, and when the young birds push their heads inside their parents' mouths they are able to suck up this unusual form of sustenance.

For juvenile mammals, of course, true milk is the mainstay of parental feeding, an evolutionary step that put the emphasis on maternal rather than paternal care. During their earliest days, sucking at the nipple becomes the centre of life for infant mammals. Different species have slightly different systems of arranging mealtimes. In some, such as dogs, there is no set order. It is a case of first come, first served, at the nearest nipple available. But with sharp-clawed species such as cats there is too much risk of scratching the mother if there is any squabbling or scrabbling for position. In these species, each infant develops an attachment to one particular nipple and makes straight for it when the mother lies down. Each kitten can identify its own nipple by its individual smell – the odour having been imparted to it by its own kittenish slurping on previous occasions.

With piglets there is also a set arrangement with personal nipple-stations for each individual, but here the reason is slightly different. In this case the nipples nearest the mother's head provide the best milk supply and the first piglets born in a litter make for these instinctively. As each piglet arrives on the scene it heads for the nipple as near to the front of the mother's body as previous occupancy will permit. Once all the young are in position, that is the sequence they will always adopt at future mealtimes. They are so reliable in their positioning that it is possible to put letters on their backs which will spell out a word every time they feed.

Most mothers will only suckle their own young but some mammals are so communal that they will permit any infants to take milk from them at feeding time. This is true of the Cape hunting dog, one of the pack-hunters of Africa, where a single female can be seen standing patiently in the colony with a whole swarm of pups clamouring to suck beneath her. After the pups have grown older they are also weaned in a communal manner. The adults return from their hunt and regurgitate half-digested meat to all the young, regardless of their particular parentage. Such extreme sharing is rare, however.

When the young need protection they may hide while the parents try to draw the attention of their enemies away from the spot where they are

Many small birds, such as this song thrush, keep their nests clean by removing the droppings of their young. The droppings are contained in a gelatinous sac that facilitates their removal, enabling the parent birds to carry them away from the nest and dispose of them at a distance.

Young nestlings, such as these greenfinches, have a gaping response to the arrival at the nest of an adult bird. The sight of the gaping mouth is a powerful stimulus to the parents to provide food, and parent birds are kept busy with almost non-stop journeys in search of titbits for the rapidly growing young.

The parents of some birds swallow the food they are collecting for their nestlings and start to digest it before giving it to them. When they arrive back at the nest, they regurgitate it in a form that is easier for the nestlings to accept. This method of feeding is common among seabirds such as albatrosses (top). Some mammals also regurgitate food for their young. This adult African hunting dog (above) brings back half-digested meat from the kill and presents it to the pups at the communal den. The food is shared between all the pups, including those that are not her own. Pigeons (below) are unique in providing a special crop-milk for their nestlings. It is a liquid that originates from the swollen lining of their crops.

concealed. Another parental device is to transport the young to safety. Carrying the offspring on the body is a form of caring that can be found even in some invertebrate species, such as scorpions and certain spiders. Females of these animals may be seen with their backs completely swarming with young. As scorpions are born, their mother carefully twists her pincers to provide them with an easy climb up. The young clamber slowly up the claws and then they hold on to her back as tightly as they can with their own tiny pincers. Packed closely together they ride with her for several days and if they are accidentally knocked to the ground she stops and searches for them. When she finds them she waits while they climb back up again before setting off once more.

Many species of fish carry their young in the mouth as a protective device. At the first sign of danger, the parental jaws gape open and the small fry dash for cover. The jaws snap shut and remain so until the danger is past, when they open once more and the young emerge to forage again.

Hardly any birds appear to carry their young, but sightings are claimed for both nightjars and woodcocks, where chicks are said to have been held tightly between the parental feet during flight. The more common method of removing a brood from one place to another is to call to them and persuade them to follow. Ducks can often be seen travelling in a single file, with the mother in front and the chicks behind. This following response is based on early learning called imprinting. Whatever the ducklings see when they first hatch becomes 'mother' as far as they are concerned, even if it happens to be a human being, a dog, or an orange balloon. Just so long as it is a large moving object, they will accept it as the 'parent to be followed'. For one famous Austrian naturalist this proved something of a problem. Having imprinted a group of ducklings, he found that they followed him closely wherever he went. On one occasion he lay down to sleep in the long grass, assuming they were somewhere near by. When he awoke and sat up he could see them nowhere, but when he turned round he found that they had been following him so very closely that he had squashed them all flat and their corpses were lying there in a neat row on the spot where he had slept so peacefully. Such are the hazards of having the wrong parent.

Mammals carry their young in several ways. The most popular method is that seen in cats, dogs and many other species. The infant is picked up by the scruff of the neck and transported in the parent's jaws. The special infantile response to this treatment is to lie very still and never to struggle. If a den is disturbed, a mother cat will find a new location and then carry her kittens or cubs there one by one. This process reveals that cats cannot count, because they alway go back for one extra trip, just to make sure that nobody has been left behind.

She-bears adopt a carrying method that looks extremely hazardous. They transport their young by taking the whole of the cub's head in the mouth. The rest of its body dangles down from the huge jaws, but the grip is so finely judged that no harm comes to the infant. Some rodents carry their young by grabbing the skin not of the neck but of the belly. Young shrews may travel in a spectacular caravan, each one gripping the tail of the one in front, with the foremost youngster hanging on to its mother's tail.

Anteaters often carry their young on their bodies when moving about. The giant anteater's single offspring travels on its back; the pangolin's rides on its tail. The baby sloth rides on its mother's underside, as does the young sea otter. Some wood rats and voles have a more primitive method of transportation. When the mother flees from an enemy, the young simply cling on tight to her nipples and bump and thump along the ground beneath her. Although this would appear to be risky, their teeth and her nipples are well adapted to this mode of escape, and no blood is drawn.

According to old etchings and drawings, the female opossum always carries her young hanging from her tail. They are depicted with their small prehensile tails twined around hers, but in reality this appears to be the exception rather than the rule. They do, however, frequently travel on her back and this must cause her considerable discomfort, as they continue to take rides in this manner even when they are quite large.

Similar discomfort must be experienced by many of the pouched marsupials, such as kangaroos and wallabies, because their young are all too eager to return to the snug protection of the pouch even when they are almost as big as their parents. It is surprising that the mothers do not send them packing long before this stage is reached, but they seem to be extremely tolerant of their cumbersome offspring.

With primates the basic method of travelling is to cling as tightly as possible to the mother's fur. This position is held even when the parents are taking great leaps through the trees. In quiet moments the babies are allowed to experiment by crawling along the branches themselves, but the mothers are ever-watchful and grab them up at the slightest sign of risk or danger. Some monkeys have been seen to bend branches together to help their young pass from one tree to the next and have occasionally been observed to make a bridge of their own bodies across which their infants can pass.

Many avian and mammalian parents are extremely solicitous over the problem of temperature control, taking great care to prevent their offspring from becoming too hot or too cold. Some ground-nesting birds that live in hot regions have been known to carry a mouthful of water back to their scorching chicks and pour it over their bodies as a cooling device. Other parent birds have been seen, their wings carefully spread to create shade at the nest, protecting their nestlings from the direct sunlight, even though they themselves are clearly in great distress from the baking heat.

For fish the problem is more often one of providing enough oxygen. Nesting species often spend many hours fanning a current of water over their nests, beating vigorously with their pectoral fins. Others gargle their young in their mouths as a way of increasing their oxgyen supply.

All mammals provide liquid nourishment in the form of milk for their offspring. The act of suckling not only feeds the young but also helps to strengthen the bond of attachment between mother and baby because of the intimate body contact involved. The milk is a complete food for the growing offspring, but varies in richness from species to species. The fat content of milk in monkeys, such as these macaques (above), is very low – less than 3 per cent. In man it is nearly 4 per cent, and in felines, such as this lioness (above, left) it is over 6 per cent. But the richest milk of all is that of the seals – the harp seal's milk has a fat content of nearly 43 per cent.

Some hot-country birds, such as this African shoebill, carry water in their mouths and dribble it to their thirsty young.

Scorpions protect their young by carrying them on their backs (right). The mother will defend them from enemies by threatening with her sting-tipped tail. After a few days the young jump off and fend for themselves. Some spiders also provide travel facilities for their young (above). These are rare examples of maternal care among invertebrate animals.

Many mammals, such as the Tamandua anteater (opposite, top), and the olive baboon (opposite, bottom), also transport their young on their backs, where they cling tightly to the parental fur.

Most frogs deposit their spawn in water and abandon the eggs to their fate, but some carry them on their backs. When they hatch into tadpoles, they may remain clinging to their parents' backs, as do the young of this forest frog from Trinidad.

It is often said that animal parents 'teach' their young, but this is a slight distortion. Occurrence of direct teaching has not been proved, but parents do encourage their growing young to accompany them and watch them as they forage, hunt, drink, hide, flee, or perform other essential adult behaviour patterns. The young learn by example and occasionally they are punished if they appear to be indulging in dangerous activities. This is close to teaching, if not the real thing, and it certainly helps to equip the young animals for a more successful adulthood.

One special kind of learning takes place during infancy that is of crucial importance for successful breeding, and this is the seemingly simple process of infants coming to recognise their own parents. As already mentioned, imprinting takes place rapidly during the early stages of the parent-offspring relationship. During a brief, sensitive period, the parent and offspring become irreversibly fixated on one another. This fixation demands intimacy, and separation causes acute distress.

For a close relationship of this type to work, there must be precise identification of the individuals involved. It is not simply a matter of learning to recognise a member of your own species, it is a question of establishing specific, personal characteristics that pin down the parent or offspring as *yours* rather than someone else's. As a young animal spends time with its parent, it becomes more familiar not only with those sights, sounds and smells that are typical of all members of its species but also with those that are unique to the individual. Both these learning processes are essential for later life. Careful studies of swans have demonstrated how they operate.

It was discovered that young swans had certain beak details that were very like those of their parents. The dark patches on the beaks of the parental pair were always rather similar in shape to those on the beaks of their own cygnets. It was also observed that when these young cygnets grew up and selected mates for themselves they tended to choose partners that had rather different markings. They did not choose the wrong species, of course – merely members of their own species that looked less like their parents than might have been expected by chance. This result shows that there is a degree of outbreeding imposed upon the new adults by virtue of their special infantile learning. This ensures that the species avoids the dangers of too much inbreeding.

(Opposite, top left) At birth, marsupials are little more than embryos and they must reach the pouch and attach themselves firmly to one of their mother's nipples if they are to survive. They may stay permanently attached to their nipple for up to two months, before releasing it and becoming more mobile.

(Above) Ducks and geese must rely on their young to follow them when they move about. They are ill-equipped for transporting their offspring in any other way. But the ducklings and goslings are extremely attentive and follow closely behind the parent in an orderly line.

In species where the young cannot ride on the mother's body, she must resort to some other method of carrying them. Rats (opposite, top right) and lions (opposite, bottom) carry their offspring gently but firmly in their jaws when moving them from one site to another. The young respond to the parental neck-grasp by going limp.

Parental fish have a special problem – ensuring that their eggs are well oxygenated. Many achieve this by fanning the clutch of eggs with their fins, passing a current of fresh water over them. This jewel cichlid (right) is clearly exerting itself to keep its clutch fresh.

In other words, when young animals learn what their parents look like, in minute detail, they store this memory and then in adult life respond sexually to individuals that look approximately, but *not closely*, like their own parents. Imprinting is therefore a double process – part positive and part negative. The positive part makes sure that the young stay with their parents throughout infancy, and the negative part makes sure that, when they mature, they do not mate with close kin. In human terms we refer to this as an incest taboo and imagine that it is something invented culturally by us. But in reality it goes much deeper and is something we share with many other species.

A final point on parental behaviour concerns the attitude of adults towards infants that are not their own. In most cases strange young are ignored or attacked, even if their parents have been killed and they are starving or in distress. In a few instances 'aunties' will help out and may feed young that are not their own, but only usually in close-knit social groups. The typical response of a new male coming into a group after the death or removal of the old male is to ignore or even kill and eat the old male's offspring. In this way no time is wasted helping to promote the older animal's genes. All efforts now go into perpetuating the genetic properties of the new male.

An intriguing exception to this rule has been noticed in a species of Japanese anemone-fish. If a male parent in the process of looking after a clutch of eggs is killed or deposed by a new male, the newcomer takes over the paternal duties and devotes as much energy to rearing the old male's clutch of eggs as if they were his own. Genetically this does not seem to make sense, but further investigation has revealed that if he does not do this the mate of the removed male will not tolerate him. Unless he acts as a foster-parent, she attacks him and refuses to mate with him. Only when he starts caring for the young does she stop butting him and beating him up. His reaction is to rear the old clutch of eggs and then to mate with her and rear his own clutch. It is a short-term altruism that gains a long-term advantage, but it is a system that only works in species where the females are unusually aggressive.

The greatest incubating task in the world of birds is that faced by the ostrich (opposite, top). Here a male tries unsuccessfully to cover the huge clutch of eggs laid by several females. The ones on the outside are not properly incubated and are also most vulnerable to predation. It has recently been discovered that the females, some of which help with incubation and some of which do not, can recognise their own eggs. The incubating females make sure that their own eggs are well placed in the middle of the nest and will move the eggs of non-incubating females to the outside. The advantage of allowing non-incubating females to add eggs to the huge clutch is that these peripheral eggs help to protect the inner ones by being available for the occasional hungry egg-thief.

(Opposite, bottom) If the sun is too hot, a parent bird, here a Bonelli's eagle, may protect its vulnerable, downy offspring by shading them with its outstretched wings.

The young need a great deal of rest and their parents must be prepared to rest with them and keep a constant watch for any signs of danger. Even the large predators, such as these cheetahs, must take precautions where young cubs are concerned. Small cheetah cubs may easily fall prey to hunting lions and the mothers often have difficulty in defending their slow-moving offspring.

Play Behaviour

Acrobatic play: here a young cheetah (below) tests its arboreal skills and an infant chimpanzee experiences the excitement of branch-swinging (opposite).

PLAY BEHAVIOUR is restricted to the higher forms of life. It does not occur in invertebrates, fish, amphibians or reptiles and it is rare in birds. Reports of playful fish have been based on misunderstandings. Some fish have been seen leaping over pieces of floating wood in what appears to be a gymnastic exercise, but in reality they are attempting to remove skin parasites. Only among mammals has playful behaviour been observed as a common activity and even there complex play is restricted largely to carnivores and primates. The play of hoofed animals usually involves no more than bursts of erratic fleeing. Carnivores and primates on the other hand indulge in a variety of play actions and often do so for very long periods of time each day, during their juvenile period.

There are three main types of play: acrobatic, investigatory and social. In acrobatic play the animal throws itself into wildly abandoned gymnastic movements during which it utilises all its muscles. Leaping, jumping, gliding, sliding, rolling, twisting, rotating, running and even somersaulting are performed at varying and often mounting intensities. A young animal first tries out a particular kind of athletic movement and then repeats it. Then it repeats it again, and again, until, often by accident, some variation creeps in. Now this variation is incorporated, making a new type of leap or twist, and gradually a new form of acrobatics grows out of an old one. The animal pushes this to the limit until, perhaps, it hurts itself and pauses. Then it switches to a different pattern of play and explores that.

Several important qualities of play are evident here: risk-taking, energy expenditure, the establishing of a theme and the varying of that theme. Throughout the play-bouts the young animal is edging itself towards the outer limits of its muscular abilities, learning how far it can go and where its weaknesses lie.

In investigatory play, objects found in the proximity of the young animal are examined and tested. They may be grasped and let go, bitten, hit, rolled, juggled, balanced, thrown and chased, and caught and 'killed'. Even if they are edible they are not eaten. Their value lies in their role as a source of inventive manipulation.

In social play, exaggerated interactions occur with companion animals. Fleeing from companions, chasing them, fighting them, mating with them, hunting them and killing them are all performed in an unrealistic manner. The fighting does not hurt, the mating does not involve real sexual acts, the predation is only mock-killing. The young animals learn a great deal about the feel of the bodies of their companions, both siblings and parents, and once again test the limits of their own bodies when engaged in vigorous physical activity.

The difference between playful behaviour in all these contexts and non-playful behaviour is that play is uneconomical in its execution. The elements of playful behaviour are exaggerated, amplified, re-ordered, repeated and often fragmented, when compared with non-playful behaviour. But why should it happen like this? Why should animals expend their energy in this way? The usual answer is that, when young, they need to practise their adult activities, to rehearse for the tough life ahead of them. Play is envisaged as a training programme for the strengths and skills they will require in adult life. But it has been argued that extravagant playfulness is not necessary for this. When a human being is practising the violin, he does not do so in an exaggerated, playful way. He tries to get it right. His actions are serious and controlled, not wildly abandoned. In a similar manner, young animals could set about serious, controlled attempts to carry out adult actions and could, in that way, prepare themselves for adult life, when survival becomes a daily challenge. So what is the special secret of play that makes it so important to perform

it in a way that is so different from other patterns of behaviour?

The answer, it seems, is that the playfulness of play is a signal to other members of the group, advertising the fact that the actions being performed are not to be interpreted in the usual manner. In other words, the young animals *are* going through a training process for later life, but they are making sure that their actions are not being misinterpreted. By frolicking, frisking and gambolling they transmit the clear message that 'this is only a rehearsal'. In this way they can put their parents at ease. Without an exaggerated, romping element in their actions, their mothers might imagine that they really were fleeing from a predator, or were wrestling a rival in a serious fight. This would alarm them immediately and they would intervene in some protective role. But the special quality of the play movements is so characteristic that even we, belonging to a completely different species, can understand them.

It is not only parents that must understand, but also the companions with whom social play takes place. It must be made abundantly clear to them that the attack which is about to be launched is a mock-assault and not the real thing. The capering, prancing movements make this clear once play is under way, but something extra is needed – some kind of 'let's play' signal is required to initiate the encounter.

Play-signals differ from species to species and are highly characteristic. In the human species we use the smile and laughter. Chimpanzees have a special play-face, with the mouth open but the lips drawn over the teeth to hide them. In dogs, play is invited with a crouching body, the front legs fully bent and the rear legs straight, so that the head end of the body is close to the ground while the rump is high in the air. Coyotes use a strange, squirming roll. Mongooses whip their tail. Foxes make little running movements with their hind legs, while standing on the spot. Polecats make stiff-legged jumps. Badgers perform little head-twists. Brown bears also use a strange head movement, a kind of 'head-wobbling'.

In some of the most playful species there is more than one play-signal. In

Investigatory play: a young leopard explores the strangeness of a moving 'rock' (above, left). Many predatory actions occur in playful form in the investigations of juvenile carnivores.

A young macaque, (above) from the safety of its mother's embrace, explores the flexibility and anatomy of its own hands.

Dogs offer a highly characteristic play invitation signal: the play bow. The front legs are bent and the head and the front part of the body held low, while the rump is held high. This signal is directed towards companions in an attempt to persuade them to play. In origin it is a fleeing intention movement and the message of the posture is 'If you will chase me I will run away.'

addition to making the play-bow, a dog may also attempt to initiate play by offering a play-object to a companion, or by lying down and rolling on its back. The red fox is said to have no fewer than twenty-two play-signals, but why it should need so many is something of a mystery.

It is perhaps significant that those species which show most intense and frequent play are the ones that live the most opportunistic lives. Monkeys, apes, dogs and foxes – these are the animals that are ever-active as adults, and one of the reasons why they play so much more than other animals may well be that they have evolved nervous systems that abhor inactivity. When matters of survival are well taken care of, these animals will still want to remain busy and active. As juveniles their survival is ensured by their parents, who feed them, clean them and protect them. This is perhaps a reason why play is so common in young animals. It would follow from this that if adult animals are also well cared for, they too should show some degree of playfulness. And that is precisly what we observe with our domestic dogs and cats, where even elderly adults may still engage in an occasional bout of play. In a sense, such animals are still juveniles in their relation with their human owners, who are their pseudo-parents, feeding them and protecting them from harm. So their playfulness is quite appropriate to their social role.

Significantly, the only group of birds that have been seen to play in a convincing manner are the crows, especially the ravens. These are opportunists *par excellence*, with the highest intelligence of all avian species. Young ravens have been seen to indulge in balancing games, deliberately seeking out the thinnest and most difficult branches on which to stand. They also engage in play-fighting, with rapid role-reversal of a kind that is exclusive to playful encounters. In real fights winners do not suddenly switch to behaving like losers and then back again, but playing animals do this all the time, first one being the attacker and then the other.

Social play: young cheetahs (below, left) and meerkats (below) play-fight. In bouts of mock fighting the role of chaser and chased keeps changing in a way that would never happen in a serious dispute. The excitement in social play is in experiencing as wide a range of roles and actions as possible, with a conspicuous lack of muscular economy.

Needless to say, the human animal, being the most intelligent and opportunistic species in existence, is also the most playful. And, since this intelligence has given us a modern lifestyle in which survival is no longer a day-to-day problem, our playfulness has stretched more and more into the adult phase. The childlike adult has become the norm and amazing extensions of playfulness are all around us. Many of our social activities are mature forms of adult play. We go play-eating in our restaurants and play-drinking in our bars and pubs. As in all play, it is not the basic survival activities that are important but their playful exaggeration and ornamentation. Our playful curiosity has been elaborated into our arts and our scientific research. Play-hunting has become sport, acrobatic play has become dancing and gymnastics. Play-fighting has become politics. The names have been changed to make us feel more sophisticated but the playful roots of our greatest preoccupations are clear enough to any objective animal-watcher.

Parents spend a considerable time playing with their offspring. Some moments are gentle and tender, as between this orang-utan mother and her baby (above left), and some are rough and painful, as these lions demonstrate (below). Parental restraint in the face of boisterous offspring is often remarkable.

Cleaning Behaviour

Animals are often misunderstood. When we use insults such as 'filthy beast' or 'dirty animal' to describe an unpleasant human being, the implication is that to be an animal is to be unclean. This is a gross distortion. Wild animals cannot afford to be unclean. They would quickly become inefficient or diseased. In reality they are always striving to be immaculately clean and perform a great deal of grooming and preening behaviour every day. The only truly dirty animals are those unfortunate enough to have been kept in appalling conditions by human owners, and it is presumably those that have given rise to the widely used insults.

There are a few apparent exceptions to this general rule, but on closer inspection they can easily be explained. Some species love to wallow in mud-pools, covering themselves in a thick layer of mud that dries and cakes on their skin, making them look extremely grubby and bedraggled. Pigs and rhinos are particularly partial to this seemingly slovenly activity, but it has a special significance in their lives. Far from making them dirty, it is in fact making them clean, because it acts as an anti-parasite device, suffocating pests that have already burrowed into their skin and creating a mud-barrier for insects either biting or attempting to lay eggs on their large bodies. It also, incidentally, helps to keep the wallowers cool, not only during the submersion in the mud but also during the water-evaporation that follows as they dry out.

Pigs and rhinos are not alone in this behaviour. Mud-bathing is also a popular pursuit of water-buffaloes, deer and a variety of other large mammals, all of which benefit in a similar way. Red deer stags, especially, devote considerable time to wallowing in spring and autumn. They seek out a suitable spot, preferably in a peat bog area, and there they paw at the ground with their front feet until they have churned up the peat into a 'soiling bath'. Then they roll and rub themselves in it, dislodging a great deal of dead hair and defeating the dangerous parasites that plague them in these seasons.

Another mammalian action that at first sight, or first smell, gives the impression of being a deliberate 'dirtying' is the dung-rolling of wolves and dogs, which sometimes occurs when they encounter the droppings of other animals as they patrol their territories. But again this has a special significance. Wolves use the added odour as a mask of their own scent. In this way they can camouflage themselves for the hunt.

Apart from these special cases, there is a strong urge in wild animals to keep their body surfaces well groomed and in lustrous condition, free of all foreign matter. They achieve this by regular bouts of cleaning involving often complex sequences of movements. The most elaborate forms of this behaviour are seen in birds, rodents and primates, but some kind of toilet behaviour is found even in invertebrates. Insects pay particular attention to their antennae and the surfaces of their compound eyes. The supposedly grimy house-fly spends a great deal of time cleansing its head, body and wings when it is at rest. Bees must repeatedly rid themselves of pollen after they have been visiting flowers and they possess an intricately designed notch and comb on their front legs with which they can stroke clean their antennae. Pollen on the body is removed and collected on their hind legs on special storage organs called pollen baskets.

Amphibians can do little more than wipe their eyes and their mouths with their front feet, and the cleaning repertoire of reptiles is much the same, although they do engage in some vigorous rubbing and pushing against hard objects when they are in the act of shedding their skins. A snake begins by rubbing the side of its face against a stone or wood surface. If the old skin cracks open around the lips it can then move forward through the skin-sheath it is discarding. If all goes well, it can advance like a finger being pulled out of a tight glove, leaving the old skin inside out as it

Animals are often called 'dirty' but in reality they are extremely clean. Even the apparently messy actions of wallowing in mud-pools are essentially cleansing techniques that rid animals of their skin-parasites and provide them with a caked mud barrier against further insect attacks. This is a common activity of African buffaloes and elephants (below), and hippopotamuses and warthogs (overleaf).

Reptiles such as snakes (left) and geckos (below) acquire clean skin by the dramatic process of shedding. The skin of geckos fragments into large flakes, but that of snakes is pulled off like a sheath.

slithers away, gleaming and bright in its new colours. If the old skin does not break away cleanly, the reptile may have to continue rubbing and writhing against rocks or logs for several days until finally the last clinging patches have been torn away.

Because of their dramatic method of replacing old skins with new ones, reptiles have a considerable skin-care advantage over other animals. Birds and mammals, by contrast, have serious problems. Feathers and hairs demand special attention.

The cleaning sequence of a small bird is fascinating to watch. When a little finch flies down to a pool of water and drinks, it will look around to check for dangers and will then, cautiously, start to bathe. If it is uninterrupted, the behaviour pattern that follows may well last for more than half an hour, during which time over a thousand separate cleaning actions will be performed in a special, predictable sequence. Here is a typical sequence, much simplified:

1. The bird flies to the edge of the water.
2. It drinks and looks around.
3. It drinks again and then gives a head-shake with the head out of water.
4. It hops into the water and hesitates for a moment.
5. Standing in the shallow water, it leans forward and performs a head-shake with its head under the surface of the water. Then it straightens up.
6. Now it bends its legs, lowering itself into the water and at the same time raises its ventral feathers fully. It also lowers its tail, so that it is submerged in the water.
7. It dips its head under the water, shaking it as it does so. At the same time it shakes its wings and sometimes its tail.
8. It raises its head up out of the water and shakes its wings again. The tail is spread and shaken too.
9. It now gains momentum with its bathing actions, alternating movements 7 and 8: head-down-shake/head-up-shake/head-down-shake, and so on.
10. The bird hops out of the water and shakes itself completely several times.

The bathing behaviour of birds follows a long and complex sequence. This Australian crimson chat responds to the sudden arrival of rain puddles by initiating such a sequence with vigorous body shakes.

11 The bird flies heavily up to a branch.
12 It wipes its bill repeatedly on the branch, cleaning both sides.
13 It rubs its face on the branch, removing water from around the eyes.
14 It shakes its head several times and wipes its bill again.
15 It shakes its body several times.
16 It twists its head round and presses the tip of its bill on to the top of its rump, just at the point where the tail meets the body. There it has a small oil gland and quickly squeezes a drop of oil on to the surface of its bill.
17 It then scratches its bill vigorously, distributing the oil all over its surface.
18 It starts to preen itself, swiping its bill through its plumage, moving from one part of the body to another.
19 It oils its beak again and scratches once more.
20 It returns to preening again.
21 It suddenly stops preening and performs a vigorous wing-whirring action, as if flying-on-the-spot. This is a powerful shaking movement but it is quite different from the 'shuffling' shake used earlier. Its function is clearly that of drying out the wings. It is accompanied by a similar fast vibration of the tail.
22 Now the bird makes a few preening movements again and alternates these with wing-whirring. This is the long period of cleaning, with the initially short preening bouts growing gradually longer and longer between the wing-whirrings.
23 Finally, the bird performs some very long and uninterrupted preening bouts and with these its full cleaning is complete.

This pattern varies slightly from species to species but the version described here clearly shows the complexity of bird cleaning. Its function is to remove dirt and parasites and to re-oil the feathers. It is also vital that each feather should be well smoothed and that it should lie neatly against neighbouring feathers to form an efficient insulating layer. Birds have a much higher temperature than we do and they must trap a layer of air within their plumage if they are to maintain their body heat. If the feathers are dishevelled they cannot do this and they quickly suffer the consequences.

An additional function of cleaning is that it provides the bird with much needed Vitamin D. When the oil that is spread on the plumage with each cleaning is exposed to sunlight, the vitamin is synthesised and can then be absorbed orally with the next bout of preening. Birds that have suffered damage to their oil glands have been observed to develop the vitamin-deficiency disease rickets. But the main function of the oil is undoubtedly to provide waterproofing for the feathers. This is borne out by the fact that it is the aquatic species of birds that possess the largest oil glands.

Some birds, such as chickens, prefer a dry bath to a wet one. They perform the same actions as the water-bathers, but they use a dusty dip in the ground rather than a pool. The ubiquitous house sparrow, famed for its opportunism in all things, is happy to bathe in either water or earth, according to which is most readily available. Dust-bathing and sand-bathing seem to be just as efficient at cleansing the plumage as water-bathing, with the feather-parasites being thrown off with the solid granules as they are shaken away by the vigorous body movements.

In addition to water, dust and sand, some birds also bathe in ants, smoke, flames and the sun. Ant-bathing has fascinated observers because when engaged in it the birds appear to be in a state of intoxication. They give the impression of being avian junkies enjoying a 'fix'. When anting, the birds flop out on a teeming ant colony, ruffle their feathers, spread their

Aquatic birds have a special repertoire of bathing actions, employed while afloat. These include wing-thrashing, seen here in a swan (above).

Many species of birds engage in dust-bathing, making a dip in the ground, crouching in it and then shuffling their feathers vigorously. Gallinaceous birds such as partridges (below) and jungle fowl (bottom) are particularly fond of this type of plumage cleaning.

wings and encourage the angry ants to crawl all over them. Some birds, in addition, take up individual ants in the tip of the bill and wipe them along their feathers, especially the tips of their wing feathers. The suggestion is that the reward obtained from this strange behaviour is the tingling sensation of heat produced by the pungent formic acid given off by the ants. Some observers see this as a form of direct skin stimulation similar to that experienced by men using a strong after-shave lotion. Others see it more as an ant-sniffing or ant-tasting addiction. The more orthodox view holds that, far from being some kind of avian acid trip, the birds are in reality applying formic acid to their plumage as a method of ridding themselves of irritating skin parasites. The dramatically increased acidity of their skin surface is supposed to be intensely noxious to the parasites and to drive them away or kill them.

The anting behaviour of the jay has fascinated observers. The bird settles on an ant nest, ruffles its feathers and allows the ants to crawl into its plumage. It also takes individual ants in its bill and wipes them through its wing feathers. According to one theory, birds use the formic acid from the ants to fumigate the feathers and rid themselves of parasites.

The final word on anting has yet to be written and the subject is a fascinating one to study. One authority has compared it with smoke bathing – an even odder pursuit. Some birds, especially members of the crow family, have been seen sitting in ecstasy on the tops of smoky chimneys, spreading their feathers and bathing in the rising smoke. Some tame birds even tried to bathe in flames, once the burning straw they were sitting on ignited. These observations have been used to support the theory that all these forms of bathing, in ants, smoke and flames, are concerned primarily with seeking heat stimulation rather than with cleansing. Opponents argue that smoke and flames would be just as efficient as formic acid in disturbing and killing skin parasites. To them all three forms of bathing are cases of 'fumigation'.

Sun-bathing gives rise to a similar argument. Many birds can be seen lying out on the grass on a sunny day, dangerously oblivious of the world around them, just like human sunbathers on a beach. They spread their wings, ruffle their feathers and tilt their heads, often closing their eyes. In this condition they are easily approached by predators, and the risk involved in the activity must mean that it has some very special counterbalancing value to the birds. What this is remains a subject of debate. Again, on one side are those who see it as a hedonistic heat-seeking addiction and on the other are those who see it more prosaically as yet another case of feather-maintenance. The intense heat of the sun is thought to drive away parasites, which retreat in panic into the earth beneath the reclining bodies of their hosts. When the birds get up and leave the spot, they leave their parasites behind. In addition, the heat of the sun's rays is thought to act as a tonic for the feathers and undeniably helps to convert the oil on them into Vitamin D.

Like birds, mammals spend much time attending to the condition of their body surfaces. Rodents spend long periods going through a fur-cleaning routine every bit as complex and predictable as the preening of birds. A session usually begins with mouth-washing, the paws being licked and then rubbed over the lips. The rubbing movements then expand to take in the whiskers and the rest of the face. Next, the forelegs are wiped up and over the ears. At this point the animal rapidly grooms the length of one flank with both paws. Then the other flank is treated in the same way. From here, the next region tackled is the hind legs, and finally the tail is cleaned. As with birds there are many minor variations, but each species follows a set pattern that appears to be completely inborn for its kind. A rodent with large back feet can be seen holding one foot in its front paws while cleaning each of its back toes individually. Those with large ears often pay special attention to the ear canals, moistening the tips of their rear toes and then inserting them into each ear with extreme delicacy, after which they withdraw them, lick them, and then insert them again.

Many species of rodents enjoy sand-bathing, but rubbing the body in sand is not always what it seems. In some species it is primarily a scent-

Sun bathing is common and widespread in reptiles, such as this terrapin (top), mammals, such as these meerkats (middle) and this sea lion (bottom), and birds. Birds adopt a variety of special sunning positions, including the Spread-wing Posture, with both wings opened and held sideways, seen here in maribou (opposite, top), the Spread-eagled Posture, with both wings opened, tail fully spread and all body feathers ruffled, seen here in a blackbird (opposite, bottom left) and the Loose-wing Posture, in which the body is held vertically and the wings are lowered and opened to expose their undersides to the sun, seen here in a goliath heron (opposite, bottom right).

Like birds, rodents such as this brown rat (right) engage in a long and complicated cleaning routine, grooming their fur with great care. Primates also spend much time delicately examining and cleaning their fur (below). They do not employ the rapid, vigorous sequence typical of rodents. Instead they use their hands to work slowly through the fur, section by section, removing any dirt or dead skin they find.

depositing action and caution is needed when interpreting this behaviour as a toilet activity.

Primates, with their flat nails, spend hours searching through their fur for dirt, parasites and dead skin. They appear to eat almost anything they can find in their fur, smacking their lips as they do so. Some of the lower primates such as bushbabies and their relatives have retained a special 'grooming claw' on the second toe of each hind foot. All the other toes bear flat nails, but these two sharp, pointed claws enable the animals to indulge in a good, dog-like scratch when they feel a skin irritation.

Hand-searching is not the mammal's only grooming technique. Licking the fur is popular with many species, especially felines. Dogs lick their paws but seldom their coats. Most mammals will nibble an irritation with their teeth and large ones may try to rub it away by scraping against trees, rocks or branches. Some try rolling on their backs on the ground, while others resort to swishing their long tails or twitching their ears to dislodge pests. Many have the ability to shiver a small section of skin if an insect lands on it, making it take off again in panic. As with birds, vigorous body-shaking is common in mammals, especially when they are emerging from water, as any owner of a large dog will know to his cost.

With both birds and mammals there are frequent cases of one animal cleaning the body of another. Birds have a special invitation posture in which the head is tilted back and the neck feathers ruffled. This has preening-appeal to a neighbouring bird, who finds the sight of the raised feathers irresistible and starts to clean them. In all such cases, the areas cleaned by a companion are just those that the owner itself cannot reach with its mouth.

It has been argued that social preening of this type (called 'allopreening' to distinguish it from 'autopreening') is not primarily concerned with feather maintenance but is basically an aggression-reducing tactic, converting serious head-pecking into harmless preen-pecking and defusing a tense encounter. There is no doubt that this plays a part, but it is not the whole explanation. True, relationships between paired birds and between socially bonded primates all become more friendly through this activity, and it certainly helps to create a non-hostile atmosphere, but there is also a great deal of serious body-cleaning taking place. This is especially so in primates, where the long hours spent grooming one another's fur does, inevitably, have an important cleansing effect. The groomer in such cases can stare closely at patches of skin that the owner cannot see clearly. And there is one supreme example of mutual skin-care that was clearly a cleaning act and nothing else. Two adult chimpanzees had shared an enclosure for many years and their relationship was fiery and often aggressive, but on one occasion the female of the pair approached the male, whimpering. She sat down and he immediately went to her, took her head gently but firmly in his left hand and started to pull down her left lower eyelid with his right forefinger. Peering intently at her left eye, he searched her eyelid and eventually located a small cinder that had lodged there. With extreme delicacy he removed it. The actions differed hardly at all from those that would have been used between a man and a woman under similar circumstances. On other occasions apes have been observed to remove splinters from the skin of their companions by carefully pinching with two forefingers.

Animals may lack all our wonderful soaps and powders and lotions, but dirty they are not.

As any cat-owner with ripped furniture will know, all felines like to tear at surfaces with their front claws. This has always been thought of as an act of claw-sharpening but in reality it is a combination of muscle exercise and claw renewal. The action of dragging the claws through the semi-hard surface helps to strengthen the all-important muscles that operate claw retraction and protrusion and it also assists in ripping off old, worn claw-sheaths to reveal the sharp new claws underneath. In addition it enables the cats to deposit their personal scent on the torn surfaces, from glands on the undersides of the paws, as this ocelot is doing.

Kangaroos, with their strangely proportioned limbs, indulge in a great deal of forelimb scratching especially in the region of the pouch.

Sleeping Behaviour

Although we take it for granted, sleep is one of the most mysterious of all animal behaviour patterns. And watching the way animals sleep makes it even more difficult to understand.

Consider the facts. We human beings sleep for an average of eight hours each night, during which time light sleep alternates with deep sleep. During deep sleep there are periods of vivid dreaming. These interludes, called active sleeping, occur at roughly ninety-minute intervals, giving us five dreaming sessions on a typical night. We awake feeling refreshed and ready to start a new day.

Because of this pattern we generally think of sleep as a recuperative process in which we 'recharge our batteries', but it is never quite clear what this recharging might be in precise physiological terms. One suggestion is that it is the brain rather than the muscles that needs an off-duty period, and that dreaming is a reflection of this. The dream images are said to be our faint awareness of the sorting-out that is taking place in our memory banks and brain cells as we slumber. It is as if there is a great deal of filing to be done at the end of the day, and pieces of information that are difficult to classify and file away keep coming back as awkward dream sequences. But eventually the long sleeping period, with outside stimulation switched off, allows all the mental tidying away to be completed and we are then ready to wake up and start the new day. If we only wanted to restore the condition of our tired muscles, we could do so simply by resting – lying down, fully awake, and relaxing our bodies completely. If we blot out the world with sleep it must be because of the extra demand for brain-rest.

That is the orthodox view of the function of sleep, but how does it measure up to the knowledge we have gained about the ways in which different animal species slumber? Not at all well, is the short answer. For a start, different species do not sleep for the same amount of time. The sloth lives up to its name, with 20 hours of sleep a day. Armadillos and bats sleep for almost as long – about 19 hours a day. Hamsters and squirrels curl up and slumber for 14 hours, rats and mice for 13. Chimpanzees and rabbits need 10 hours, guinea pigs and cows only 7. A horse makes do with only 5 hours and a giraffe with 4, while the ever-active shrew does not sleep at all.

These few examples from animal sleep studies create an immediate problem for the recuperation theory. Why should bats and shrews, both

In Africa, the quelea roosts (below) are so dense that sometimes whole trees collapse under the weight. Starlings (below, right) also roost in huge numbers – sometimes hundreds of thousands – and like the crowd at a football match they keep one another warm simply by their combined body heat, even if they are not touching one another. The starling roost is also an important information exchange centre, where hungry birds can attach themselves to well fed ones and follow them to good feeding grounds when they wake in the morning. It is also a protection against predation. The well fed ones settle in the safer centre of the communal roost. The weaker ones take up positions on the perimeter. This means that the outsiders are more vulnerable, but in exchange they get the 'feeding news' from the fat birds in the centre.

Chimpanzees make a fresh bed each night before settling down to sleep. They never use the same one twice, which avoids dirt or infestation. The sleeping platform only takes a few minutes to make, branches being pulled in and held down to create a springy mattress. Infants sleep with their mother until they are three, when they make their own small beds near to hers. Adult apes never sleep together, even if they are good friends.

At night certain species of marine parrot fish secrete a mucus envelope around themselves. It acts as a protective barrier. The fish completely cover themselves with this loose shield and this reduces the chances of nocturnal predation by moray eels. The parrot fish leave a small hole in the envelope, through which they can breathe.

with a very fast metabolism, need such different amounts of sleep? If sleep restores the body, then the busy, bustling shrew should spend a great deal of its time sleeping, like the bat. But it does not. And if sleep is essential for brain-rest – for filing away all the new input from the previous waking period – then why do the sloth and the armadillo need twice as much sleep as human beings? If input-sorting is the true function of sleep, we humans should be the greatest sleepers of the animal world, but we are nowhere near the top of the slumber-league.

The recuperation theory also runs into trouble if the quality of animal sleep is examined. The human pattern, with repeated bouts of active, dreaming sleep throughout the night, is found only in the higher mammals and the birds. It is absent in the monotremes – the platypus and the echidna, the lowest of the mammals – and also in all reptiles, amphibians, fish and invertebrates. In other words, cats and camels can dream, and so can parrots and pigeons, but snakes and salamanders cannot, and nor can minnows or moths. So the idea that we sleep in order to dream and indulge in input-sorting cannot be applied to all species. If reptiles can do without it, why do birds need it? Again, the recuperation theory of sleep provides no answer. So we must look for some other explanation. But if recovery from the exhaustions of the day is not the function of sleep, then what could it possibly be?

One thing is clear: sleep is not the absence of behaviour, it is a quite specific, positive behaviour pattern. It occurs in a definite rhythm each twenty-four hours and this rhythm varies from species to species. Some animals slumber through the day, others through the night. Still others sleep in the middle of the day and the middle of the night, becoming active only at dawn and dusk. These three types of animal are described respectively as diurnal, nocturnal and crepuscular. Going to sleep involves appropriate appetitive behaviour – introductory activities such as bed-making, roosting, seeking out a safe place, retreating to a den or nest, and adopting a characteristic sleeping posture. Throughout the sleep period there are frequent position-shifts that prevent cramping or numbing of muscles. Both before sleeping and after waking there are patterns of stretching and yawning.

The most characteristic feature of sleep that distinguishes it from ordinary resting is that sensory input is so dramatically reduced that slumbering animals can be approached and even caught before waking up. In the safety of our locked houses and apartments we worry little about this, and fall asleep without any anxiety about our impending vulnerability. But for a wild animal the situation is very different. Despite the safe nest or den, the risk of 'switching off' the outside world is enormous. All the usual defence mechanisms are inactivated and prowling predators could easily strike before the victim would have time to react. So if sleep has this huge disadvantage it must have some very powerful advantage to counterbalance it.

An intriguing new theory suggests that the true function of sleep is not recuperation but inactivation. Sleep, this theory says, is the 'great immobiliser'. Sleep switches animals off because they are better served if for part of each twenty-four hours they are *incapable* of action. They might be vulnerable when sleeping, but they would be even more vulnerable if they were awake.

If there is part of the day or night when animals do not have to be active because all their needs are satisfied, or when they cannot be active because it is too dark or too bright for them, or too hot or too cold, then it is better that they should be 'knocked out' by sleep rather than that they should sit around fretting about their inactivity, or be leaping up hazardously at every slight disturbance.

It is argued that only species which must keep feeding almost non-stop

(Above) Elephants sleep very little, the average number of hours of recumbent slumber per night in captive elephants being only two and a half. Before lying down they gather together grasses, twigs and leaves to make soft pillows for their heads and flanks.

(Above, right) Giraffes sleep lying down, but with the neck erect for most of the time. They do go into a deep sleep for brief periods, with their long necks wrapped around them and with their heads resting on their rumps. This deep sleep amounts to no more than twenty minutes every twenty-four hours.

to stay alive, or those that are intensely vulnerable to predation throughout the day and night, would benefit from staying awake all the time, or for most of it. The recuperation theory would demand that all such species should have a decent period of sleep every twenty-four hours, but this is not what we find. Indeed, the facts fit the immobilisation theory much more closely, with vulnerable creatures such as antelopes sleeping hardly at all. Hoofed animals in general, with their heavy bodies forcing them to take ground-level rest periods in the open, are very brief sleepers, taking only a few hours at the most each night. An oddity here is the elephant which also sleeps very little, averaging about $2\frac{1}{4}$ hours a night. This giant beast has little to fear from predators, but it does have another problem – its huge weight. If it sleeps too long on its side, its internal organs suffer, and it too, therefore, must reduce its slumber periods to a minimum.

Many prey species cannot afford to sleep for very long, but their predators are less restricted. The big cats, such as lions (opposite) and leopards (right), spend about two-thirds of their lives sprawled out in relaxed slumber.

If sleep is simply an evolutionary device for keeping animals out of harm's way at difficult times – more of an anaesthetic than a rest-cure – then why is it that sleep deprivation in humans causes such problems? Experiments in which individuals are deliberately deprived of sleep cause them considerable suffering as the sleepless days go by. This is seen as irrefutable support for the recuperation theory. If the subjects lie down and rest, it is not enough. They need the complete switch-off or they will suffer mental anguish. The inactivation theory argues that these artificial tests are merely frustrating a powerful, inborn drive – the drive to fall asleep – and that it is this frustration and not the long period of waking that is causing the damage. It is also pointed out that certain exceptional individual humans can manage well with little or no sleep and do not suffer from this extreme régime any more than non-sleeping mammals such as shrews.

The way in which sleep has evolved provides further support for the immobilisation theory. If we turn the clock back to the time when reptiles dominated the earth, before mammals and birds appeared on the scene, all forms of animal life were what is commonly called 'cold-blooded'. That is, they had a variable body temperature, growing warmer in the heat of the day and cooler at night. This meant that they were all more sluggish in the cold nights and experienced a period of inertia during this phase of the twenty-four-hour cycle. Their inertia, we know today, contained no interludes of deep, dreaming sleep. It was merely a light sleep that was little more than a temporary torpor.

If any new group of animals was going to challenge the supremacy of the great reptiles, it would have to break this pattern. The mammals did it in one way and the birds in another. The earliest mammals were all small, scurrying, nocturnal creatures. Their secret was that they had managed to evolve a constant, high body temperature. Being 'warm-blooded' at night, they did not fall into the reptilian torpor and could busy themselves when the giants were slumbering. All the primitive mammals were of this nocturnal type and there are many such mammals even today. Day-active mammals have all developed more recently, filling the gaps left when the giant reptiles mysteriously departed from the scene.

The problem for the ancestral mammals was that they were at risk during the day, when the reptiles heated up in the sun and became active. What was needed was a form of sleeping that could be switched on in the daytime, despite the heat and light of the sun. This would keep them tucked safely away in their dens and nests, conserving their energies until the safety of nightfall. The more advanced, active sleeping was the answer to this problem, with dreaming perhaps introduced as a way of preventing too great a switching-off of the brain during the phases of deepest sleep and most extreme muscle relaxation. In other words, dreaming was simply a way of keeping the brain 'turning over' during an otherwise extreme state of immobilisation.

The advanced slumber pattern of these new mammals enabled them to stay asleep even during the most intrusive moments of sunny daylight. The newly evolving birds had a different problem. For them, with flight through the air as a novel solution to the challenge of predation, there was also a need for a high, constant body temperature. Without it, sustained flying would not be possible. But flying also demanded daylight. The nocturnal specialists such as owls were a long way off in the future. The earliest birds therefore had to rest at night, unlike the early mammals. But their body temperatures now remained high during this phase, as part of their increased physiological efficiency. So, unlike the torpid reptiles, these new birds had to have some way of restraining themselves from plunging into disastrous flight in pitch darkness. If they were disturbed at night, they had to avoid panic, and for them too there was therefore a powerful need for a more advanced form of sleep. Like the mammals, they therefore evolved the deep sleep pattern, with massive relaxation of muscles combined with a protective turning-over of the brain. For two quite different reasons, both birds and mammals became 'warm-blooded dreamers', immobilised for their own safety by superior sleeping behaviour.

This is the controversial new theory of sleep, but it will take some time for it to sweep away the more orthodox concept that it is a primarily restorative process. Because *we* feel restored by our slumbers we will be loath to let the old theory go, despite the fact that the animal evidence cannot be fully explained by it. Perhaps in the end a third theory will emerge that combines the two existing ones. Clearly there is much more to discover about this subject.

A final word is needed on the topic of hibernation. For some mammals, ordinary sleep is not enough. Living in regions where there is intense cold in the winter and little food, an even greater immobiliser is needed to protect them from activity that could easily prove fatal. In this even more extreme form of sleep, the metabolic processes are slowed right down until the hibernating animal is virtually in a state of suspended animation. Many small mammals are capable of dropping their body temperatures to only a few degrees above that of the environment and of reducing their energy output by as much as seventy times. Larger mammals such as bears cannot go this far, being capable of only a 50 per cent reduction in their metabolic rate.

In a sense, hibernation is a retrograde step, returning these mammals to the torpid condition of their ancient reptilian ancestors in periods of cold. But at least during their summers they can enjoy the full rewards of the warm-blooded way of life, with activities carried on independently of the more minor fluctuations between day and night temperatures.

(Opposite) Yawning, performed here by a jaguar, a lion and a gorilla, is associated with body stretching and appears to be primarily a jaw-stretching mechanism.

(Left) The diet of koalas is so poor that they seem to need an unusually large amount of sleep, curling up in the branches and snoozing away eighteen hours out of every twenty-four.

The body of a hibernating mammal, such as this tightly curled up dormouse (below), can be as much as 90 degrees Fahrenheit below its normal waking body temperature, but it can raise its temperature back to normal in only three hours. Some dormice, in very cold climates, may hibernate for as long as nine months in the year

As with so many aspects of animal behaviour, we still have a great deal to learn even about such an apparently simple act as going to sleep. Most of what we know has been learned from laboratory studies, where conditions are highly artificial and often misleading. Observations of animals sleeping in their natural habitats are few and far between and we need much more animal-watching on this subject in the years ahead, before we can fully understand this strange pattern of behaviour, a pattern which robs the average human being of one-third of his or her entire life. As already mentioned, some exceptional human individuals require little or no sleep, but at present we do not know why. If animal studies could give us the answer, perhaps one day we will all be able to lengthen our waking lives.

If this happens, we will have yet another debt to the animals with which we share this small planet. But for true animal-watchers this should always be a secondary consideration. Our primary concern should be to enjoy animals for *their* sake and not for our own. In the past we have all too often used our knowledge to exploit them rather than to celebrate them. The time has come to change that, and the better we become at animal-*watching*, pure and simple, the easier this will be.

Some birds, such as flamingos, are vulnerable at night and frequently sleep with one eye open. It has been claimed that such birds allow one hemisphere of their brains to sleep at a time, changing the open, alert eye and the alert hemisphere from time to time during the period of slumber. They can then awake refreshed, having switched off their brains without ever once being completely unwary.

References

WHY DOES THE ZEBRA HAVE STRIPES ?

Groves, C. P., 1974, *Horses, Asses and Zebras in the Wild*, David and Charles, Newton Abbot
Kingdon, J., 1984, 'The zebra's stripes', in Macdonald, D. (ed.) *The Encyclopaedia of Mammals*, pp. 486–7, Allen and Unwin, London
Mochi, U. and Carter, T. D., 1974, *Hoofed Mammals of the World*, Lutterworth Press, London

GROUPING BEHAVIOUR

Crook, J. H., 1970, *Social Behaviour in Birds and Mammals*, Academic Press, London
Tinbergen, N., 1953, *Social Behaviour in Animals*, Methuen, London

ESCAPE BEHAVIOUR

Chance, M. R. A. and Russell, W. M. S., 1959, 'Protean displays: a form of allaesthetic behaviour', *Proc. Zool. Soc. London* 132, pp. 65–70
Edmunds, M., 1974, *Defence in Animals*, Longman, London
Humphries, D. A. and Driver, P. M., 1971, 'Protean defence by prey animals', *Oecologiia* 5, pp. 285–302
Lopez, B. H., 1978, *Of Wolves and Men*, Dent, London
Tinbergen, N., 1948, *The Study of Instinct*, Oxford University Press, Oxford

PROTECTIVE ARMOUR

Edmunds, M., 1974, *Defence in Animals*, Longman, London
Hanney, P. W., 1975, *Rodents, their Lives and Habits*, David and Charles, Newton Abbot
Hoogland, R., Morris, D. and Tinbergen, N., 1956, 'The spines of sticklebacks as means of defence against predators', *Behaviour* 10, pp. 205–36

CAMOUFLAGE

Cott, H. B., 1940, *Adaptive Coloration in Animals*, Methuen, London
Wickler, W., 1968, *Mimicry in Plants and Animals*, Weidenfeld and Nicolson, London

WARNING SIGNALS

Cott, H. B., 1940, *Adaptive Coloration in Animals*, Methuen, London
Edmunds, M., 1974, *Defence in Animals*, Longman, London
Huxley, J. S., 1934, 'Threat and warning coloration in birds, with a general discussion on the biological functions of colour', *Proc. VIII Int. Ornith. Congress*, pp. 430–55
Wickler, W., 1968, *Mimicry in Plants and Animals*, Weidenfeld and Nicolson, London

CHEMICAL DEFENCE

Caras, R. A., 1964, *Dangerous to Man*, Holt, Rinehart and Winston, New York
Evans, H. M., 1943, *Sting-fish and Seafarer*, Faber and Faber, London
Halstead, B. W., 1959, *Dangerous Marine Animals*, Cornell Maritime Press, Cambridge, Maryland
Morris, R. and Morris, D., 1965, *Men and Snakes*, Hutchinson, London
Ricciuti, E. R., 1976, *Killer Animals*, Walker, New York
Whitley, G. P., 1943, 'Poisonous and Harmful Fishes', *Bull. Council Sci. Ind. Res. Australia, Melbourne* 159

DEFLECTION DISPLAYS

Cott, H. B., 1957, *Adaptive Coloration in Animals*, Methuen, London

STARTLE DISPLAYS

Blest, D., 1957, 'The function of eyespot patterns in the Lepidoptera', *Behaviour* 11, pp. 209–56
Blest, D., 1957, 'The evolution of protective displays in the Saturnioidea and Sphingidae (Lepidoptera)', *Behaviour* 11, pp. 257–309
Cott, H. B., 1957, *Adaptive Coloration in Animals*, Methuen, London
Edmunds, M., 1974, *Defence in Animals*, Longman, London

DEATH-FEIGNING

Hartman, C. G., 1952, *Possums*, University of Texas, Austin
Matthews, L. H., 1969, *The Life of Mammals*, Weidenfeld and Nicolson, London
Mertens, R., 1960, *The World of Amphibians and Reptiles*, Harrap, London
Noble, G. K., 1931, *The Biology of the Amphibia*, McGraw-Hill, New York
Varty, K., 1967, *Reynard the Fox*, Leicester University Press, Leicester

SELF-MUTILATION

Mertens, R., 1960, *The World of Amphibians and Reptiles*, Harrap, London
Noble, G. K., 1931, *The Biology of the Amphibia*, McGraw-Hill, New York
Yonge, C. M. and Thompson, T. E., 1976, *Living Marine Molluscs*, Collins, London

DISTRACTION DISPLAYS

Armstrong, E. A., 1947, *Bird Display and Behaviour*, Drummond, London
Edmunds, M., 1974, *Defence in Animals*, Longman, London
Simmons, K. E. L., 1952, 'The nature of the predator-reactions of breeding birds', *Behaviour* 4, pp. 161–71
Simmons, K. E. L., 1955, 'The nature of the predator-reactions of waders towards humans', *Behaviour* 8, pp. 130–73
Skutch, A. F., 1954, 'The parental stratagems of birds', *Ibis* 96, pp. 544–64

MOBBING

Altmann, S. A., 1956, 'Avian mobbing behaviour and predator recognition', *Condor* 58, pp. 241–53
Edwards, G. et al., 1949, 'Reactions of some passerine birds to a stuffed cuckoo', *British Birds* 42, pp. 13–19
Hartley, P. H. T., 1950, 'An experimental analysis of interspecific recognition', *S. E. B. Symposia* 4, pp. 313–36

FOOD-FINDING

Frisch, K. von, 1967, *The Dance Language and Orientation of Bees*, Harvard University Press, Cambridge, Massachusetts
Greenewalt, C. H., 1960, *Hummingbirds*, Doubleday, New York
Owen, J., 1980, *Feeding Strategy*, Oxford University Press, Oxford

LURING BEHAVIOUR

Burgess, W. E., Axelrod, H. R. and Hunziker, R. E., 1988, *Atlas of Marine Aquarium Fishes*, T.F.H. Publications, New Jersey
Cott, H. B., 1940, *Adaptive Coloration in Animals*, Methuen, London
Wickler, W., 1968, *Mimicry in Plants and Animals*, Weidenfeld and Nicolson, London

FOOD-PREPARATION

Kawamura, S., 1962, 'The process of sub-culture propagation among Japanese Macaques', *Journal of Primatology* 2, pp. 43–60
Lyall-Watson, M., 1963, 'A critical re-examination of food "washing" behaviour in the raccoon', *Proc. Zool. Soc. London* 141, pp. 371–93
Leyhausen, P., 1979, *Cat Behaviour*, Garland Press, New York
Tinbergen, N. and Norton-Griffiths, M., 1950, 'Oystercatchers and Mussels', *British Birds* 57, pp. 64–70

FOOD-STORAGE

Ewer, R. F., 1965, 'Food burying in the African Ground Squirrel', *Z. Tierpsychol.* 22, pp. 321–27
Ewer, R. F., 1968, *Ethology of Mammals*, Logos Press, London
Lyall-Watson, M., 1964, *The Ethology of Food-hoarding in Mammals*, London University
Morris, D., 1962, 'The behaviour of the Green Acouchi, with special reference to scatter-hoarding', *Proc. Zool. Soc. London* 139, pp. 701–32

MUTUAL AID

Perry, N., 1983, *Symbiosis*, Blandford, Poole, Dorset

DRINKING BEHAVIOUR

Goodwin, D., 1970, *Pigeons and Doves of the World*, British Museum, London
Goodwin, D., 1982, *Estrildid Finches of the World*, British Museum, London
Schmidt-Nielsen, K., 1964, *Desert Animals*, Clarendon Press, Oxford

CANNIBALISM

Cloudsley-Thompson, J. L., 1965, *Animal Conflict and Adaptation*, Foulis, London
Fox, L. R., 1975, 'Cannibalism in natural populations', *Annual Review of Ecology and Systematics* 6, pp. 87–106
Fox, M. W., 1968, *Abnormal Behavior in Animals*, Saunders, Philadelphia
Troyer, W. A. and Hensel, R. J., 1962, 'Cannibalism in Brown Bear', *Animal Behaviour* 10, p. 231

TOOL-USING

Boswall, J., 1977, 'Tool-using by birds and related behaviour', *Avicultural Magazine* 83, pp. 88–97, 146–59, 220–8; 84, pp. 162–6
Ewer, R. F., 1968, *Ethology of Mammals*, Logos Press, London
Morris, D., 1954, 'The snail-eating behaviour of thrushes and blackbirds', *British Birds* 47, pp. 33–49

CONFLICT BEHAVIOUR

Bastock, M., Morris, D. and Moynihan, M., 1953, 'Some comments on conflict and thwarting in animals', *Behaviour* 6, pp. 66–84
Chance, M., 1962, 'An interpretation of some agonistic postures: the role of "cut-off" acts and postures', *Symp. Zool. Soc. London* 8, pp. 71–89
Daanje, A., 1950, 'On locomotory movements in birds and the intention movements derived from them', *Behaviour* 3, pp. 48–98
Lorenz, K., 1941, 'Comparative studies of the motor patterns of Anatinae', published in English, 1971, in *Studies in Animal and Human Behaviour* vol. 2, Methuen, London
Morris, D., 1954, 'The reproductive behaviour of the Zebra Finch, with special reference to pseudofemale behaviour and displacement activities', *Behaviour* 6, pp. 271–322
Morris, D., 1956, 'The feather postures of birds and the problem of the origin of social signals', *Behaviour* 9, pp. 75–113
Morris, D., 1977, *Manwatching: a Field Guide to Human Behaviour*, Cape, London
Morris, D., 1988, *Horsewatching*, Cape, London
Tinbergen, N., 1952, 'Derived activities: their causation, biological significance, origin and emancipation during evolution', *Quart. Rev. Biol.* 27, pp. 1–32

TYPICAL INTENSITY

Dane, B. et al., 1959, 'The form and duration of the display actions of the Goldeneye', *Behaviour* 14, pp. 265–81
Goodwin, D., 1970, *Pigeons and Doves of the World*, British Museum, London
Morris, D., 1957, 'Typical Intensity and its relation to the problem of ritualization', *Behaviour* 11, pp. 1–12
Wiley, R. H., 1973, 'The strut display of male Sage Grouse: a "fixed" action pattern', *Behaviour* 47, pp. 129–52

FACIAL EXPRESSIONS

Fox, M. W., 1971, *Behaviour of Wolves, Dogs and Related Canids*, Cape, London
Hooff, J. A. R. A. M. van, 1967, 'The facial displays of the Catarrhine monkeys and apes', in Morris, D. (ed.) *Primate Ethology*, Weidenfeld and Nicolson, London
Leyhausen, P., 1979, *Cat Behavior*, Garland, New York
Morris, D., 1986, *Catwatching*, Cape, London
Morris, D., 1986, *Dogwatching*, Cape, London
Morris, D., 1988, *Horsewatching*, Cape, London
Morris, R. and Morris, D., 1966, *Men and Apes*, Hutchinson, London
Waring, G. H., 1983, *Horse Behavior*, Noyes, New Jersey

FIGHTING BEHAVIOUR

Carthy, J. D. and Ebling, F. J., 1964, *The Natural History of Aggression*, Academic Press, London
De Waal, F., 1982, *Chimpanzee Politics*, Cape, London
Southwick, C. H., 1970, *Animal Aggression*, Van Nostrand Reinhold, New York

SUBMISSIVE BEHAVIOUR

Morris, D., 1954, 'The reproductive behaviour of the River Bullhead', *Behaviour* 7, pp. 1–31
Morris, D., 1958, *The Reproductive Behaviour of the Ten-spined Stickleback*, Brill, Leiden

COURTSHIP DISPLAYS

Armstrong, E. A., 1942, *Bird Display and Behaviour*, Drummond, London
Bastock, M., 1967, *Courtship: a Zoological Study*, Heinemann, London
Halliday, T., 1980, *Sexual Strategy*, Oxford University Press, Oxford
Morris, D., 1956, 'The function and causation of courtship ceremonies', in Grasse, P. P. (ed.) *L'Instinct*, Masson et Cie, Paris
Tinbergen, N., 1953, *Social Behaviour in Animals*, Methuen, London

ARENA DISPLAYS

Gilliard, E. T., 1969, *Birds of Paradise and Bower Birds*, Weidenfeld and Nicolson, London
Hogan-Warburg, A. J., 1966, 'Social behaviour of the Ruff', *Ardea* 54, pp. 109–229
Kruijt, J. P. et al., 1972, 'The arena system of the Black Grouse', *Proc. XV Int. Ornith. Congress*, pp. 399–423
Kruijt, J. P. and Hogan, J. A., 1967, 'Social behaviour on the lek in Black Grouse', *Ardea* 55, pp. 203–40
Marshall, A. J., 1954, *Bower Birds, their Displays and Breeding Cycles*, Clarendon Press, Oxford

MATING BEHAVIOUR

Boulière, F., 1955, *The Natural History of Mammals*, Harrap, London
Ewer, R. F., 1968, *Ethology of Mammals*, Logos Press, London
Kelves, B., 1986, *Females of the Species*, Harvard University Press, Cambridge, Massachusetts
Morris, D., 1954, 'The reproductive behaviour of the River Bullhead', *Behaviour* 7, pp. 1–31
Sparks, J., 1978, *The Sexual Connection*, McGraw-Hill, New York
Wickler, W., 1969, *The Sexual Code*, Weidenfeld and Nicolson, London

NESTING BEHAVIOUR

Crook, J. H., 1960, 'Nest form and construction in certain West African Weaver-birds', *Ibis* 102, pp. 1–25
Frith, H. J., 1962, *The Mallee-fowl*, Angus and Robertson, London
Hancocks, D., 1973, *Master Builders of the Animal World*, Evelyn, London
Von Frisch, K., 1975, *Animal Architecture*, Hutchinson, London

PARENTAL CARE

Bateson, P., 1979, 'How do sensitive periods arise and what are they for?' *Animal Behaviour* 27, pp. 470–86
Bateson, P., Lotwick, W. and Scott, D. K., 1980, 'Similarities between the faces of parents and offspring in Bewick's swan and the differences between mates', *J. Zool. London* 191, pp. 61–74
Rheingold, H. L., 1963, *Maternal Behaviour in Mammals*, Wiley, New York
Ridley, M., 1978, 'Paternal care', *Animal Behaviour* 26, pp. 904–32
Yanagisawa, Y. and Ochi, H., 1986, 'Step-fathering in the anemone-fish *Amphiprion clarkii*: a removal study', *Animal Behaviour* 34, pp. 1769–80

PLAY BEHAVIOUR

Aldis, O., 1975, *Play Fighting*, Academic Press, New York
Bruner, J. S., Jolly, A. and Sylva, K., 1976, *Play: Its Role in Development and Evolution*, Penguin Books, London
Groos, K., 1898, *The Play of Animals*, Chapman and Hall, London
Loizos, C., 1967, 'Play behaviour in higher primates: a review', in Morris, D. (ed.) *Primate Ethology*, Weidenfeld and Nicolson, London
Millar, S., 1968, *The Psychology of Play*, Pelican, London

CLEANING BEHAVIOUR

Buettner-Janusch, J. and Andrew, R. J., 1962, 'The use of incisors by Primates in grooming', *Amer. J. Phys. Anthropol.* 20, pp. 127–9
Burton, M., 1959, *Phoenix Reborn*, Hutchinson, London
Eisenberg, J. F., 1963, 'A comparative study of sandbathing in heteromyid rodents', *Behaviour* 22, pp. 16–23
Harrison, C. J. O., 1965, 'Allopreening as agonistic behaviour', *Behaviour* 24, pp. 161–209
McKinney, F., 1965, 'The comfort movements of Anatinae', *Behaviour* 25, pp. 120–220
Miles, W. R., 1963, 'Chimpanzee behaviour: removal of foreign body from companion's eye', *Proc. Nat. Acad. Sci.* 49, pp. 840–3
Morris, D., 1956, 'The feather postures of birds and the problem of the origin of social signals', *Behaviour* 9, pp. 75–113
Rhijn, J. G. van., 1977, 'The patterning of preening and other comfort behaviour in a herring gull', *Behaviour* 63, pp. 71–109
Simmons, K. E. L., 1966, 'Anting and the problem of self-stimulation', *J. Zool. London* 149, pp. 145–62
Simmons, K. E. L., 1986, *The Sunning Behaviour of Birds*, Ornithological Club, Bristol

SLEEPING BEHAVIOUR

Hediger, H., 1955, *Animals Asleep*, Geigy, Basle
McFarland, D., 1989, 'Sleep as an ethological problem', in *Problems of Animal Behaviour*, Longman, Harlow, Essex
Meddis, R., 1975, 'On the function of sleep', *Animal Behaviour* 23, pp. 676–91

Index